T0134927

The Arc of Life

Grazyna Jasienska • Diana S. Sherry
Donna J. Holmes
Editors

The Arc of Life

Evolution and Health Across the Life Course

 Springer

Editors
Grazyna Jasienska
Department of Environmental Health
Jagiellonian University Medical College
Krakow, Poland

Donna J. Holmes
Department of Biological Sciences and
 WWAMI Medical Education Program
University of Idaho
Moscow, ID, USA

Diana S. Sherry
School of Communication
Institute of Liberal Arts
 and Interdisciplinary Studies
Emerson College
Boston, MA, USA

Department of Human Evolutionary
 Biology
Harvard University
Cambridge, MA, USA

ISBN 978-1-4939-8161-8   ISBN 978-1-4939-4038-7   (eBook)
DOI 10.1007/978-1-4939-4038-7

© Springer Science+Business Media New York 2017
Softcover reprint of the hardcover 1st edition 2017
This work is subject to copyright. All rights are reserved by the Publisher, whether the whole or part of the material is concerned, specifically the rights of translation, reprinting, reuse of illustrations, recitation, broadcasting, reproduction on microfilms or in any other physical way, and transmission or information storage and retrieval, electronic adaptation, computer software, or by similar or dissimilar methodology now known or hereafter developed.
The use of general descriptive names, registered names, trademarks, service marks, etc. in this publication does not imply, even in the absence of a specific statement, that such names are exempt from the relevant protective laws and regulations and therefore free for general use.
The publisher, the authors and the editors are safe to assume that the advice and information in this book are believed to be true and accurate at the date of publication. Neither the publisher nor the authors or the editors give a warranty, express or implied, with respect to the material contained herein or for any errors or omissions that may have been made.

Printed on acid-free paper

This Springer imprint is published by Springer Nature
The registered company is Springer Science+Business Media LLC
The registered company address is: 233 Spring Street, New York, NY 10013, USA

Contents

Contributors

Richard G. Bribiescas Department of Anthropology, Yale University, New Haven, CT, USA

Erin E. Burke Department of Anthropology, Yale University, New Haven, CT, USA

Hidemi Nagao Peck Department of Anthropology, Indiana University Bloomington, Bloomington, IN, USA

Ruby L. Fried Department of Anthropology, Institute for Policy Research, Northwestern University, Evanston, IL, USA

Peter B. Gray Department of Anthropology, University of Nevada, Las Vegas, NV, USA

Donna J. Holmes Department of Biological Sciences and WWAMI Medical Education Program, University of Idaho, Moscow, ID, USA

Grazyna Jasienska Department of Environmental Health, Jagiellonian University Medical College, Krakow, Poland

Rufus A. Johnstone Department of Zoology, University of Cambridge, Cambridge, UK

Karen L. Kramer Department of Anthropology, University of Utah, Salt Lake City, UT, USA

Christopher W. Kuzawa Department of Anthropology, Institute for Policy Research, Northwestern University, Evanston, IL, USA

Matthew H. McIntyre Department of Anthropology, University of Central Florida, Orlando, FL, USA

Michael P. Muehlenbein Department of Anthropology, Indiana University Bloomington, Bloomington, IN, USA

Sean P. Prall Department of Anthropology, Indiana University Bloomington, Bloomington, IN, USA

Diana S. Sherry School of Communication, Institute of Liberal Arts and Interdisciplinary Studies, Emerson College, Boston, MA, USA

Department of Human Evolutionary Biology, Harvard University, Cambridge, MA, USA

Lynnette Leidy Sievert Department of Anthropology, University of Massachusetts Amherst, Amherst, MA, USA

Jonathan Wells Childhood Nutrition Research Centre, UCL Institute of Child Health, London, UK

Chapter 1
Introduction: Evolutionary Medicine and Life History Theory

Donna J. Holmes and Grazyna Jasienska

Abstract Evolutionary medicine is a developing specialty in the biological sciences. Modern Darwinian theory posits that natural selection shapes variation in anatomy, physiology, genetics, and behavior. In this volume, we examine human health across the life span with a primary focus on life history theory - an analytical framework for understanding how organisms allocate resources to maximize reproductive success. From the standpoint of life history theory, individuals face genetic and physiological trade-offs to optimize investment in reproductive and other priorities at different stages of the life course. Trade-offs can be reflected in variation in nutritional and social status, fertility, disease risk, mortality - and other parameters conventionally thought of as "health" variables. In evolutionary terms, reproduction is the most important capacity organisms possess, since it is the means for passing genes on to the next generation. For long-lived primates, like humans, adaption to a given environment is reflected in the number of healthy, potentially reproductive offspring produced. Reproduction is costly, and timing of reproduction can have major effects on the health and mortality of both sexes over the life course. This volume brings together intellectual perspectives of biological anthropologists and evolutionary biologists from developmental biology, reproductive ecology and physiology, demography, immunology, and the biology of aging. Our aim is to showcase diverse ways in which emerging styles of evolutionary analysis can enrich our perspectives on medicine and public health. In this first chapter, we introduce basic concepts from human evolutionary biology and bioanthropology that are used throughout the volume.

To Peter T. Ellison: with appreciation from colleagues, collaborators, and former students

D.J. Holmes (✉)
Department of Biological Sciences and WWAMI Medical Education Program, University of Idaho, MS 3051, Moscow, ID, 83844-3051, USA
e-mail: djholmes@wsu.edu

G. Jasienska
Department of Environmental Health, Jagiellonian University Medical College, Grzegorzecka 20, 31-531 Krakow, Poland
e-mail: jasienska@post.harvard.edu

© Springer Science+Business Media New York 2017
G. Jasienska et al. (eds.), *The Arc of Life*, DOI 10.1007/978-1-4939-4038-7_1

1.1 Introduction

Evolutionary medicine and evolutionary public health are rapidly developing areas of biological science. (Williams and Nesse 1991; Nesse and Williams 1996; Stearns and Koella 2008; Jasienska 2013; Stearns and Medzhitov 2015) Although the term "Darwinian medicine" (later replaced by "evolutionary medicine") was first proposed in 1991 by George Williams and Randolph Nesse, an evolutionary approach to human health began much earlier in the history of science. Charles Darwin's grandfather, Erasmus, is cited as one of the first physicians to propose in the eighteenth century a connection between nature and human health. Trevathan and colleagues (1999) have provided an excellent review of the history of evolutionary medicine.

Long before the field of evolutionary medicine became recognized, many human biologists and biological anthropologists based their ideas and explanations of research findings on the principles of evolutionary theory (Ellison 2003). One of the most dominant avenues of research in the field of human biology has been the study of human variation and an emphasis on the richness of differences in anatomy, physiology, genetics, and behavior among individuals. These differences are not classified as either "normal" or "pathological" as they are in the medical field, but instead are viewed as the result of evolutionary processes.

The Arc of Life Evolution and Health over the Life Course showcases ways in which research conducted by biological anthropologists can enrich our understanding of variation in human health. This book looks at human health from the evolutionary perspective with particular focus on life history theory (Stearns 1992). (Roff 2002) Life history theory, a part of evolutionary theory, provides an explanatory framework for understanding how organisms allocate their time and energy in ways that maximize reproductive success. Given that resources are limited, individuals face trade-offs in terms of allocating resources to different stages of the life course, such as when to enter the stage of reproductive maturation or embark upon the first birth. *The Arc of Life* examines the consequences of life history trade-offs, a fundamental principle of evolutionary biology, on various aspects of human health in both sexes across the life cycle.

1.2 Brief Overview of Life History Theory

In evolutionary terms, the ability to reproduce is the most important capacity living organisms possess, allowing for passing genes on to the next generations. For animals that reproduce more than once, like many mammals, reproductive effort involves investment in multiple breeding episodes over the life course. For long-lived primates, like humans, successful adaption to a given set of environmental circumstances is measured in terms of lifetime reproductive success or the number of healthy, potentially reproductive offspring produced relative to the success of other individuals in the population.

Reproduction is costly for humans and other long-lived primates. Not only does reproductive effort involve pregnancy and lactation, it may also include long-term provisioning of mates or nurturing offspring long past weaning. A variety of factors contribute to lifetime reproductive success in humans and nonhuman primates, including the timing of birth, maturation, and spacing of reproductive episodes, the amount of energy invested in procuring mates and producing offspring, and the ener-getic expenditures and risks involved in sustaining pregnancy, producing milk, and other forms of offspring nurture and care. The resources needed for reproduction may be in short supply or available only seasonally. Longer reproductive life spans also necessitate maintaining healthy bodies and nonreproductive structures between breeding periods over the long haul, as well as obtaining adequate nutrition to sup-port reproductive organs and offspring during the demands of breeding. Complex social networks and learning systems often must be negotiated or mastered. Reproductive individuals can incur significant risks in the form of predation, compe-tition with conspecifics, physiological stress, exposure to diseases or accident, or the loss of energy stores vital for future health and reproductive investment.

The trajectory of the life span—the timing of maturation, births, and deaths—is referred to by evolutionary biologists as the *life history*. Rates of birth and mortality (*vital rates*), in turn, are referred to as *life history traits*. Evolutionary theory predicts that these demographic variables represent key aspects of an animal's phenotype that have been shaped by natural selection and contribute to lifetime reproductive success, just as other key phenotypic characteristics contribute to an individual's evolutionary success relative to others in a population. Natural selection is expected to produce organisms with the most successful adaptive "strategies" or complex sets of trade-offs among all of the benefits, costs, and risks associated with reproduction and associated phenomena. Life history trade-offs can be extremely complex and difficult to measure directly, particularly in human or other primate populations, since physiological measurements and experimental manipulations are difficult to carry out.

Some of the physiological structures, hormones, and growth factors necessary for reproduction in mammals—mammary glands, gonads, and sex steroid hormones like estrogens, progesterone, and testosterone—carry special physiological risks (e.g., an increased risk of cancer), as well as providing obvious reproductive bene-fits. Traits that provide developmental or reproductive benefits early in life may become liabilities later on and may contribute to the aging process (Finch and Rose 1995; Finch 2007; Cohen and Holmes 2014; Williams 1957). While natural selec-tion is expected to optimize reproductive traits and factors during the reproductive life span, no direct evolutionary benefits accrue by surviving or remaining healthy past the period in which reproduction is possible. Traits are only adaptive insofar as they contribute to lifetime reproductive success, synonymous with Darwinian fit-ness. Variation in the human reproductive life span itself, therefore, reflects past natural selection as well as shorter-term responses to evolutionary change in response to novel environments.

1.3   Human and Primate Life History and Reproductive Strategies

In general, primates are relatively large, long-lived mammals with big brains and complex social behavior. Humans are longer lived than our closest primate relatives, the chimpanzees and gorillas, and other mammals of similar body size. Modern humans in industrialized countries can now live up to 30 years or more past the peak reproductive years. Humans also have a long reproductive life span characterized by large investment in each offspring, with females generally investing more than males. As in the great apes, reproduction is costly energetically and pregnancy and childbirth relatively risky. Females must make trade-offs between the number of offspring produced and the quality of care provided. For example, women with high parity often have poorer health and shorter life spans.

Investment in an unsuccessful reproductive episode is also costly in terms of lost alternative reproductive opportunities. The seasonal timing of reproduction can be critical in terms of the availability of mates or food, as can the timing in terms of the life course: very young and very old women, for example, are not as successful on average at conceiving or bringing a pregnancy to term. Males, on the other hand, may incur their biggest challenges in establishing mating relationships and obtaining access to fertile, receptive females.

Primate reproductive patterns are expected to vary with environmental conditions and population variables (including population density). Reproductive variation can take the form of adaptive behavioral, developmental, or physiological responses to environmental cues (*phenotypic plasticity*)—all limited by the genetic constraints of the population. When environmental variation exceeds the ability of animals to respond adaptively within phenotypic norms, natural selection (in the form of mortality or reproductive failure) will act to produce changes in gene frequencies and, with these, new phenotypic norms.

1.4   Cultural and Epidemiological Shift

Over the past two centuries, cultural and technological change has occurred more rapidly than at any previous time in the history of the human species (Boyd and Silk 2009). Since the Industrial Revolution in the mid-1800s, mortality rates at all phases of the human life cycle have declined. In industrialized countries like the United States, people are living on average 30 years longer than they did in 1900. With the use of reliable contraceptives, childbirth and parenting are often delayed until significantly later in life, and fertility rates are much lower than ever before.

These modern shifts in the human life history pattern are primarily the result of cultural, rather than evolutionary, change. People today live longer than our ancestors because modern sewage disposal, clean water, and the widespread availability of medical care lessen the susceptibility to infectious disease and death from accidents. Famine and warfare are less widespread now than other times in human history. Most people do not have to worry about dying of tetanus, an attack by wild animals, or exposure to the elements. While some significant genetic and phenotypic change accompanying biological evolution has undoubtedly occurred during this period as well, cultural change has generally outpaced the rate of evolutionary change.

1.5 Health in Evolutionary Perspective

Natural selection has shaped human anatomy and physiology to maximize reproductive success or Darwinian fitness, not to maximize health (or "physical fitness") over the entire life course. From the evolutionary point of view, it is important for an organism to be healthy only insofar as good health contributes to an increased ability to pass genes on to the next generations. Any trait that increases reproductive success will be promoted by natural selection even if having the trait contributes to faster aging or poor health at older ages. The following example illustrates the principle: good nutrition for girls during childhood leads to an earlier age at menarche, even though an earlier age at menarche is a significant health risk. Women who mature relatively early have a higher risk of breast cancer. Nonetheless, they also have a longer reproductive life span, which may lead to higher lifetime reproductive success. In adult women, good nutrition also leads to increased levels of reproductive hormones. Although this is beneficial for fertility, lifetime exposure to high levels of reproductive hormones increases the risk of breast cancer. Similarly, in adult men, good nutrition leads to increased levels of the reproductive hormone testosterone. Again, this is beneficial for mate competition and reproductive success, but also increases the risk of prostate cancer. Furthermore, a single gene often encodes multiple traits, which may have antagonistic effects on health. Such antagonistic genetic pleiotropy is often responsible for good health and an increased ability to reproduce at a young age but, at the same time, for increased susceptibility to health problems at older ages.

Evolutionary medicine points out that the human body is vulnerable to health problems such as chronic disease (heart disease, cancer, and diabetes) and diverse medical conditions such as myopia, skeletal and joint degeneration, and infertility for two main reasons: (1) the "mismatch" between ancient genes and modern environments given the rapid acquisition of novel diets and lifestyles, as well as a mismatch on a shorter timescale between the prenatal environment and the adult environment, and (2) inherent design "flaws" in the human body arising from the inevitable evolutionary trade-offs involved in maximizing lifetime reproductive success.

1.6 Contributions from Biological Anthropology

The Arc of Life Evolution and Health over the Life Course brings together biological anthropologists with expertise in the areas of developmental biology, reproductive ecology and physiology, demography, immunology, and senescence. Biological anthropologists tend to gear their research toward understanding not only the fitness consequences of life history trade-offs (infant mortality, pregnancy outcome, lifetime reproductive success) but also the physiological and sociocultural processes responsible for their regulation. Biological anthropologists also tend to rely heavily on both cross-cultural comparisons involving traditional or non-Western societies and cross-species comparisons in order to understand natural human variation, especially in the context of potential adaptation versus pathology. The collected volume presented here aims to illustrate the diverse ways in which biological anthropologists contribute to and further advance the field of evolutionary medicine and public health.

On a final note, the coeditors chose to title this book *The Arc of Life* as a tribute to Peter T. Ellison by referencing one of the chapters from *On Fertile Ground*. Most of the authors included in this volume were influenced by Peter's ideas and research over the decades. Peter Ellison's work in the area of human reproductive ecology established one of the most insightful applications of evolutionary theory to the intersections between biological anthropology, physiology, and human health. His innovative approach has extended to all stages of human life history and brought new understanding to many of the subdisciplines within the field of human evolutionary biology. Peter's work continues to inspire new generations of researchers in both the laboratory and field in ways that inform about human health.

References

Boyd R, Silk JB (2009) How humans evolved. Norton, New York

Cohen AA, Holmes DJ (2014). Aging: evolution. Elsevier reference modules in biomedical sciences. doi:10.1016/B978-0-12-801238-3.00032-5

Ellison PT (2003) On fertile ground. Harvard University Press, Cambridge

Finch CE (2007) The biology of human longevity: inflammation, nutrition, and aging in the evolution of life spans. Academic (Elsevier), Amsterdam

Finch CE, Rose MR (1995) Hormones and the physiological architecture of life history evolution. Q Rev Biol 70:1–52

Jasienska G (2013) The fragile wisdom: an evolutionary view on women's biology and health. Harvard University Press, Cambridge

Nesse RM, Williams GC (1996) Why we get sick: the new science of Darwinian medicine. Vintage Books, New York

Roff DA (2002) Life history evolution. Sinauer Associates, Sunderland

Stearns SC (1992) The evolution of life histories. Oxford University Press, New York

Stearns SC, Koella JC (eds) (2008) Evolution in health and disease. Oxford University Press, New York

Stearns SC, Medzhitov R (2015) Evolutionary medicine. Sinauer Associates, Sunderland

Trevathan WR, Mckenna JJ, Smith EO (eds) (1999) Evolutionary medicine. Oxford University Press, New York

Williams GC (1957) Pleiotropy, natural selection and the evolution of senescence. Evolution 11: 398–411

Williams GC, Nesse RM (1991) The dawn of Darwinian medicine. Q Rev Biol 70:1–22

Chapter 2
Intergenerational Memories of Past Nutritional Deprivation: The Phenotypic Inertia Model

Christopher W. Kuzawa and Ruby L. Fried

Abstract Human and animal model research shows that prenatal nutrition influences early development and has long-term effects on adult biology and chronic disease. Much of this literature has emphasized the limited maternal capacity to buffer the fetus from stressors that negatively impact development. An alternative perspective recognizes that a subset of prenatal biological responses reflect an ability to adaptively change how the body regulates metabolism, hormone production, and other biological functions in anticipation of postnatal environmental conditions. The applicability of this concept to humans has been challenged on the basis of the long duration of the human life span and the imperfect correlation between environmental conditions during early and later life. The phenotypic inertia model proposes a solution to this problem: if maternal physiology and metabolism transfer nutrients or hormones in relation to the mother's average life experience, rather than to the specific conditions experienced during that pregnancy, this could provide a more reliable basis for the fetus to adjust its long-term strategy. This hypothesis is supported by evidence that fetal nutrition is buffered against short-term fluctuations in maternal intake during pregnancy in women who are not on the extreme ends of energy balance, while showing evidence of sensitivity to a mother's early developmental and chronic nutritional experience. Maternal buffering of fetal nutrition in humans is predicted to limit the deleterious impact of nutritional stress experienced by the mother during pregnancy while also attenuating the long-term health benefits of short-term dietary supplements consumed during pregnancy. According to this model, maternal interventions aimed at improving the health of future generations via fetal nutritional programming will be most effective when they emulate sustained, rather than transient, nutritional improvement.

C.W. Kuzawa (✉) • R.L. Fried
Department of Anthropology, Institute for Policy Research,
Northwestern University, Evanston, IL 60208, USA
e-mail: kuzawa@northwestern.edu

© Springer Science+Business Media New York 2017
G. Jasienska et al. (eds.), *The Arc of Life*, DOI 10.1007/978-1-4939-4038-7_2

2.1 Introduction

There is now much evidence that constrained fetal nutrition has effects on a broad array of functional capacities and disease outcomes (Barker 1994; Barker et al. 1989; Roseboom et al. 2001, 2006). Birth weight is commonly measured and recorded and as a result has frequently served as a proxy for prenatal nutrition, hormone exposure, and other gestational conditions. Low birth weight (LBW) is a global health concern as it predicts elevated risk of cardiovascular disease and metabolic syndrome and also contributes to infant mortality and leads to deficits in indices of human capital, such as body size, strength, lung capacity, educational attainment, and even wages (Hancox et al. 2009; Levitt et al. 2000; Yliharsila et al. 2007). In addition, there is growing evidence that maternal overnutrition, as demonstrated by maternal obesity and type II (non-insulin-dependent) diabetes, leads to increased risk of cardiometabolic disease in offspring (see Armitage et al. 2008; Dörner et al. 1988). These findings linking infant and adult well-being to both under- and overnutrition during pregnancy underscore the need to understand the underlying biological processes and targets for intervention (Gluckman and Hanson 2005; Kuzawa and Thayer 2012).

From a public health or policy perspective, the finding that fetal nutrition has long-term effects on adult health and functional capacities has led to a hope that modifying the diets of pregnant women will result in benefits for future generations. Perhaps the strongest evidence for a benefit of supplementing the dietary intake of pregnant women comes from an intervention in the Gambia in which women in mid-gestation received an additional daily allotment of 1028 kcal, along with calcium and iron (Ceesay et al. 1997). This intervention yielded a sizeable 201 g increase in offspring birth weight during the hungry season, when nutritional status deteriorates due to negative energy balance secondary to food scarcity, high workloads, and high burdens of infectious disease due to the heavy rains. During the harvest season, the same intervention yielded a 94 g increase (roughly 4 oz).

The Gambian study suggests that substantial and sustained improvement in maternal dietary intake can improve fetal nutrition and growth, with the greatest benefits experienced during acute bouts of energetic and nutritional stress. Unfortunately, few other similar interventions have yielded comparable levels of success in improving birth outcomes. In a systematic review of balanced protein/calorie pregnancy supplement trials, Kramer and Kakuma (2003) found a reduced risk of giving birth to small-for-gestational age offspring. However, these macronutrient supplementation trials did not yield significant increases in offspring birth weight, length, or head circumference. As illustrative examples, increases in birth weight of 51 and 41 g, and a decrease of 40 g, were reported in three of the larger dietary supplementation trials of undernourished women in Guatemala (Mora et al. 1978), New York City (Rush et al. 1980), and India (Kardjati et al. 1988), respectively. Among the five studies that provided balanced protein/energy supplements to adequately nourished women, birth weight increased in two but declined in three, with a weighted-average nonsignificant increase of around one ounce.

These findings underscore a paradox and also a policy dilemma: *fetal* nutrition apparently has large effects on future health, but increasing *maternal* macronutrient intake during pregnancy only nudges fetal growth and birth outcomes. To the extent that fetal growth is a useful proxy for the nutrients that cross the placenta, this suggests that current intervention strategies (that generally augment a woman's dietary intake during pregnancy) are not likely to yield large improvements in fetal nutrition and downstream metabolic programming. What might account for this apparent disconnect between the foods that the mother consumes during pregnancy and the nutrients that the fetus receives across the placenta?

The present chapter integrates principles from biological anthropology and evolutionary biology to shed light on the possible origin and function of developmental responses to early nutritional environments. It has been known for over 50 years that developmental plasticity, or the capacity of developmental biology to be modified in response to the environment, is an important way by which human populations adapt to changing environments (Lasker 1969). The literature documenting long-term effects of early environments expands the concept of human developmental plasticity by pointing to pathways of communication between mother and offspring, such as nutrients and hormones passed across the placenta and via breast milk (Bateson 2001; Bateson et al. 2004; Gluckman and Hanson 2005; Kuzawa and Pike 2005). An evolutionary framework highlights the need to consider the timescale of ecological change that these flexible systems are likely built to accommodate (Kuzawa 2005). Using an evolutionary approach, we argue that finding ways to communicate cues (e.g., nutrients, hormones) intergenerationally that mimic *sustained* environmental change, in contrast to the short-term, transient changes represented by most nutritional supplementation trials, will prove key to designing interventions that improve long-term functional and health outcomes in future generations.

2.2 Intergenerational Effects and the Human Adaptability Framework

To gain insights into effective strategies for modifying early developmental plasticity to improve long-term health, it is helpful to first consider the fundamental question of why the body modifies its biology in response to early life experiences. For some outcomes, the simple explanation is that environmental stressors impair healthy development, as illustrated by the fact that fetal nutritional stress can impair the growth of organs (Harrison and Langley-Evans 2009; Nwagwu et al. 2000). This is thought to explain some of the links between smaller birth size and poor adult health, such as the fact that LBW babies tend to have smaller kidneys with fewer nephrons, which predisposes them to hypertension and renal failure later in life (Iliadou et al. 2004; Lampl et al. 2002).

While some effects of early environmental conditions result in impairment or damage, other physiological outcomes influenced by fetal nutrition are not as easily explained. Instead, they appear to result from changes in regulatory set points that influence how the body prioritizes specific functions or responds to experiences over the life course. For example, in children and adults born with a lower birth weight, weight is preferentially deposited in the form of fat accumulation in the abdominal or "visceral" region. The fat cells in this region are innervated with sympathetic nerve fibers that secrete adrenaline, thus allowing the body to rapidly mobilize a usable source of energy from these cells when faced with a stressful challenge (Hucking et al. 2003). The free fatty acids that are released by the visceral depot when the body is faced with a stressor have additional cascading biological repercussions. The very act of mobilizing reserve energy stores sends a signal to the liver that the body is under duress and that glucose should thus be spared for use in more critical functions, such as brain metabolism. This is achieved by reducing the sensitivity of tissues like muscle and liver to the effects of insulin (i.e., insulin resistance) (Kuzawa 2010). Not only do low birth weight individuals deposit more fat in this rapidly mobilized depot, but there is also evidence that their fat cells mobilize *more* stored fats when exposed to the same dose of adrenaline, making this stored fuel rapidly available for use (Boiko et al. 2005).

There is nothing about the development and function of the fat deposits of these individuals that hints at either damage or impairment. Unlike the example of impaired kidney growth, there is no shortage of cells in the affected organ or tissue. Instead, these findings suggest that prenatal nutrition can change regulatory set points to alter the priorities with which scarce resources are used within the body: in this example, the body preferentially deposits any excess energy in fat deposits that are more easily and rapidly accessible when the body is faced with stress (Boiko et al. 2005). Many of the changes that are triggered in response to prenatal conditions appear to have similar origins in altered regulation, suggesting that they are examples of adaptive adjustments in developmental biology rather than simple impairments of organ growth (Kuzawa and Thayer 2011).

2.3   Developmental Plasticity as a Means of Adaptation

The concept of adaptation is one of the organizing principles of evolutionary biology and refers broadly to changes in organismal structure, function, or behavior that improve survival or reproductive success. Genetic adaptation more specifically refers to the process by which gene variants (alleles) that code for beneficial traits become more common within a population's gene pool through the mechanism of natural selection. Although natural selection is a powerful mode of adjustment at the population level, many environmental changes occur more rapidly than can be efficiently dealt with by changes in gene frequency, which require many generations to accrue. To cope with more rapid change, human biology includes additional, more rapid adaptive processes (Kuzawa 2005; Lasker 1969). The swiftest ecological fluctuations (e.g., fasting between meals or the increase in nutrients that our bodies need when we run) are handled primarily via homeostatic systems, which respond to changes or perturbations in ways that offset, minimize, or correct deviations from an initial state. Operating not unlike a thermostat, which maintains a constant temperature by turning the furnace on and off, homeostatic systems modify physiology, behavior, and metabolism to maintain relatively constant internal conditions despite fluctuations in features like ambient temperature, dietary intake, and physical threat. The distinctive features of homeostatic systems include their rapid responsiveness, their self-correcting tendencies, and the fact that the changes they induce are reversible.

Some environmental trends are chronic enough that they are not efficiently buffered by homeostasis, but also do not persist for long enough for substantial genetic change to occur. As such, organisms may not rely on homeostasis and natural selection to adjust biological strategies to these intermediate timescale trends. A simple example illustrates how a sustained change in experience might overload the flexible capacities of a homeostatic system if this was the only means available to help the organism adapt (Bateson 1963). In this case, imagine an individual that has recently moved to high altitude where oxygen pressure is lower, resulting in an elevated heart rate that increases blood flow and thus the rate that oxygen-binding red blood cells pass through the lungs. By engaging a homeostatic system (heart rate), the body has activated a temporary fix to help compensate for the low oxygen pressure. However, this is only a short-term solution that comes with a cost: not being able to increase heart rate further if the need arises, like when fleeing from a predator. Thus, the homeostatic strategy of elevating heart rate may work for short-term acclimation, but is a poor means of coping with chronic high-altitude hypoxia.

Over longer time spent at high altitude, additional biological adjustments ease the burden on the heart, such as increasing the number of oxygen-binding red blood cells in circulation. However, individuals who grow and develop at high altitude exhibit an even better strategy for coping with hypoxia. They grow larger lungs, a developmental response that increases the lung's surface area for oxygen transfer, thus obviating the need for temporary and more costly short-term adaptations (Frisancho 1993). This change in developmental biology is an example of *developmental plasticity*, which allows organisms to adjust biological structure on timescales too rapid to be dealt with through natural selection, but too chronic to be efficiently buffered by homeostasis (Kuzawa 2005).

These mechanisms can be viewed as allowing the organism to fine-tune structure and function to match the needs imposed by their idiosyncratic behavioral patterns, nutrition, stress, and other environmental experiences that cannot be "anticipated" by the genome (West-Eberhard 2003). Unlike homeostatic changes that are transient, growth and development occur only once, and thus plasticity-induced modifications tend to be irreversible once established. In this sense, developmental plasticity is intermediate between homeostasis and natural selection in both the phenotypic durability of the response and the timescale of ecological change that it accommodates.

2.4   The Phenotypic Inertia Model

Because some of the biological changes induced by intrauterine or infancy cues appear to reflect modifications of regulatory set points rather than developmental damage, it has been speculated that some components of developmental plasticity might be initiated to help the fetus prepare for conditions likely to be experienced after birth (Bateson 2001; Gluckman and Hanson 2005; Kuzawa 2005) (see Fig. 2.1). Some of the adjustments made by the nutritionally stressed fetus in utero (such as the aforementioned tendency to deposit more abdominal body fat and the glucose-sparing effects of muscle insulin resistance) could provide the advantage of saving scarce glucose for use in more essential functions after birth (e.g., brain metabolism) if the environment remains nutritionally stressful after birth (Gluckman and Hanson 2005; Kuzawa 2010). By this reasoning then, nutrition, hormones, and other gestational stimuli experienced by the developing fetus might convey information about local ecological conditions, thereby allowing the fetus to adjust priorities in anticipation of the postnatal reality that s\he is likely to experience (Bateson 2001). Bateson (2001) describes this as the mother sending a "weather forecast" to the fetus, while Gluckman and Hanson (2005) label this a "predictive adaptive response."

   One challenge to this notion of long-term anticipatory adaptation comes from the fact that humans have long life spans. Because humans typically live many decades, any conditions experienced during a few months of early development, such as gestation or early infancy, may not serve as reliable cues of environments likely to be experienced in adult life (Kuzawa 2005; Wells 2003). We have argued that it is precisely the brief and early timing of many of the body's periods of heightened developmental sensitivity that paradoxically could *help* the developing organism overcome the challenge of reliably predicting future conditions (Kuzawa 2005; Kuzawa and Thayer 2011). The mother's physiology could buffer the fetus against the day-to-day, month-to-month, or seasonal fluctuations in the environment while passing along more integrative information on average conditions experienced by the mother in recent decades or the grandmother prior to the moth-

Ecological cycle duration		Adaptation	
Years		Mode	Process
0.00000001	seconds		
0.0001	hours	Physiologic	Homeostasis & Allostasis
0.001	days		
0.1	months		
1	years	Developmental	Plasticity
10	decades		
100	centuries	Intergenerational	Inertia
1000	millenia	Genetic	Natural selection
1000000	millions		

Fig. 2.1 The timescales of human adaptation (modified after Kuzawa 2008)

er's birth and during lactation. Because the mother's biology and behavior have been modified by her lifetime of experiences (including her own gestational environment, and thus her offspring's grandmother's experience), the nutrients, hormones, and other resources that she transfers to the fetus in utero or to her infant via breast milk could correlate with average local conditions and experiences rather than to the vagaries of what the mother happens to experience during any given week or month of gestation itself (Kuzawa 2005; Wells 2003). If maternal physiology buffered out transient, short-term nutritional variations while conveying more reliable average information, this could provide the fetus with a more useful basis for adjusting characteristics such as growth rate, body composition, and/or nutritional requirements as environmental conditions gradually shift across decades of his/her lifetime or over several generations (Kuzawa 2005).

There is accumulating evidence that the mother's body does convey average, rather than transient, ecological information to the fetus. As noted above, despite the general tendency for populations faced with low socioeconomic conditions and chronic nutritional stress to have reduced birth weights, supplementing the diets of pregnant women in these populations generally has minimal effects on the birth weight of immediate offspring (Kramer and Kakuma 2003). In contrast, studies provide evidence that a mother's own early life nutrition, or nutrition across development, may be a strong predictor of the birth weight of her future offspring. In most studies that evaluate this, fetal growth of the mother is a stronger predictor of offspring birth weight than is the father's birth weight, pointing to possible maternal effects, either indirect genetic or epigenetic (for review see Kuzawa and Eisenberg 2012). Similarly, in a pooled sample of five large cohort studies in low- and middle-income countries, both maternal and paternal birth weight and early postnatal growth were found to be significant predictors of offspring birth weight, but these relationships were more consistent (and stronger) among females, pointing to a possible intergenerational effect of the mother's developmental nutrition (Addo et al. 2015).

Leg growth is particularly sensitive to nutritional conditions during infancy and early childhood and thus provides a useful retrospective proxy of developmental conditions experienced by the mother (Frisancho et al. 2001). In the United Kingdom, a woman's adult leg length was a stronger predictor of offspring birth weight than was her trunk length (Lawlor et al. 2003). In a similar study, a woman's leg length measured during her own childhood (around 7 years of age) was the strongest maternal anthropometric predictor of her future offspring's birth weight, even after adjusting for adult size and stature (Martin et al. 2004). Most recently, a study in the Philippines found leg length to be the strongest predictor of both offspring birth weight and placental weight, whereas trunk length was only weakly related to birth weight and unrelated to placental weight (Chung and Kuzawa 2014). Because leg growth is among the most nutritionally sensitive components of stature growth, these studies suggest that a mother's own infancy or early childhood nutrition can have lingering intergenerational effects on offspring fetal growth, which likely manifests in part through alterations in placental growth and size.

Collectively, these studies, viewed alongside the relatively modest success of nutritional supplementation trials during pregnancy, suggest that long-term or chronic nutritional history may be an important influence on the resources transferred in support of fetal growth and birth size (secondarily impacting the many traits and functions that are sensitive to and "downstream" of prenatal experience), while fluctuations in the mother's intake during pregnancy itself (though important) have comparably modest effects. This *phenotypic inertia* in fetal nutrient transfer—reflecting the lingering biological but nongenetic effects of the mother's average experiences in the past—could allow the fetus to track those dynamic features of environments that are relatively stable on the timescale of decades or even several generations (see Kuzawa 2008; Kuzawa 2005).

2.5   Mechanisms of Intergenerational Phenotypic Inertia

The mechanisms linking maternal-fetal nutrient transfer to the mother's developmental or chronic nutrition are unknown, but there are interesting potential candidates. First, it has been hypothesized that modifying the mother's growth rate during early critical periods in skeletal growth (when long-term growth trajectories are set) could set the mother's "productivity" and thereby have carryover effects on offspring fetal growth once productivity is reallocated from self-growth to supporting offspring growth (Kuzawa 2007). The greater nutritional sensitivity of leg growth during this early critical period and the finding of strong relationships between leg growth and offspring birth weight are consistent with this idea.

In addition, because maternal glucose levels during pregnancy are important predictors of offspring fetal growth and birth weight (Metzger et al. 2008), any effects of early life metabolic programming that the mother experiences could have secondary impacts on her pregnancy glucose status and thus the fetal growth of offspring. Once female offspring develop into reproductive adults, these phenotypic effects could linger and even accumulate across multiple generations (Drake and Liu 2010; Kuzawa and Eisenberg 2014; Kuzawa and Sweet 2009). Epigenetic changes are likely candidate pathways for such life course, inter- or multigenerational effects (for review see Kuzawa and Eisenberg 2014). For instance, in human populations, individuals exposed to famine while in utero exhibit alterations in methylation at genes (like IGF2) that are related to glucose metabolism and cardiovascular disease risk (Heijmans et al. 2008). Studies demonstrate the transmission of environmentally-induced epigenetic changes across one or more generations in other mammals (Franklin et al. 2010; Guerrero-Bosagna and Skinner 2012). At present, little is known about how common germline epigenetic inheritance might be in humans, but intriguing evidence has emerged from the creative use of historical records. In northern Sweden, harvest yields during the grandparents' childhood predicted mortality in matched-sex grand offspring, pointing to possible multigenerational, epigenetic, and sex-linked effects of nutritional experiences (Pembrey et al. 2006).

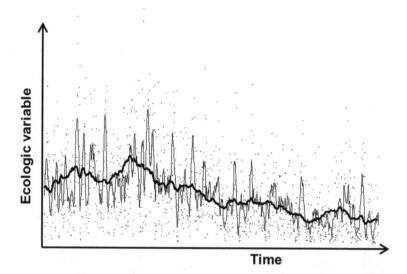

Fig. 2.2 The value of averaging as a means of identifying a trend in a noisy signal, in this case representing the availability of a hypothetical ecological resource. The two lines are running averages calculated across 10 time units (*thin line*) and 100 time units (*dark line*). As the window of averaging increases, an underlying long-term trend is uncovered. Intergenerational influences of maternal and grandmaternal nutritional history on fetal nutrition may help achieve a similar feat (from Kuzawa and Quinn 2009, with permission)

While the mechanistic specifics remain to be demonstrated, if the mother's (or father's) body or epigenome pass along biological "memories" that "sample" (and thus mirror) chronic local nutritional experiences, this could allow the fetus to make developmental adjustments in response to conditions that have dominated in recent generations and, thus, serve as a "best guess" of conditions likely to be experienced during the lifetime of the offpsring (see Fig. 2.2).

2.6 Why Do Some Nutritional Interventions Fail?

We have argued that long-lived organisms will tend to buffer or ignore transient features of their environments, but are sensitive to environmental features that are stable over longer time periods (i.e., a generation or more). We conclude by considering the implications of this idea for two research and policy domains: the biomedical use of animal experiments as models for developmental processes in humans and the design of human interventions aimed at improving long-term health.

First, what is "transient" or "stable" for individual species is inherently relative. We should expect that adaptively relevant timescales of environmental change for a human will be markedly longer than those of a rat or other short-lived species. Thus, humans should be expected to "ignore" the types of environmental changes mice or rats will modify their life trajectory in response to. After all, if a rat is born during a stressful season or year, those conditions are likely to predominate during its entire brief life. Because humans will live through hundreds of seasons, what a mouse would consider as an environmental "signal" would simply be "noise" to be buffered out by a developing human.

This perspective may help explain why individuals in the World War II Dutch Famine Winter cohort, who were exposed in utero to caloric restriction of a similar magnitude as in many animal experiments, experienced comparably small changes in birth weight followed by more modest long-term effects on metabolism and adult cardiovascular disease risk factors (Lussana et al. 2008; Painter et al. 2006) (see Fig. 2.3). Because biological processes and responses scale with traits like

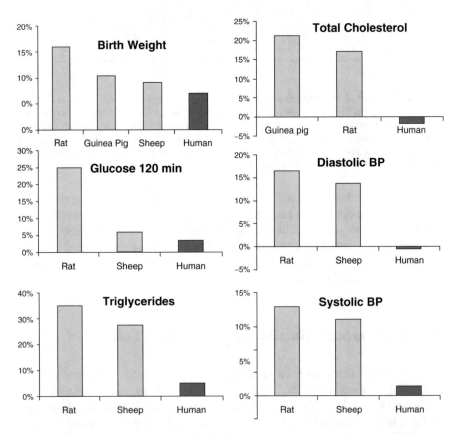

Fig. 2.3 The magnitude of change in offspring biology triggered by maternal diet restriction during pregnancy across species varying in life span (modified after Kuzawa and Thayer 2011; see original for references). All animals born to mothers who experienced caloric restriction during early gestation (30–50 % global caloric restriction). All values are calculated as the percent difference between the control group and the case group and represent averaged male and female values. All human data are from the Dutch Famine Winter adult cohort with those who experienced famine in middle or late gestation compared to the control group of those conceived after the famine (Roseboom et al. 2001)

body size and life span, mice and rats can provide a precedent for how programming might work mechanistically in humans, but remain poor guides to the magnitude of health impacts of early stressors for humans.

More specifically, the differences in findings in humans and rodents suggest that maternal nutritional stress will generally have attenuated negative impacts on offspring biology in humans, compared to other species commonly used as experimental models in biomedical research. By the same token, it seems likely that short-term improvements, as reflected in the typical design of many interventions, may yield comparably modest long-term benefits. Pregnancy supplementation trials often modify the nutritional ecology of a mother for a period of weeks or months, which represents a timescale of change that a human body should be expected to simply buffer. It would not be advantageous for a human to adjust its strategy *for life* based upon such a short-term and likely transient change in the mother's experience.

Ideally, the knowledge of the long-term benefits that accrue with favorable early life nutrition would motivate economic and health policies that improve the nutrition of entire populations to optimize health over the course of two or three human generations. In the absence of such policies, the principles of biological adaptation, and of timescale, may lead us to effective shortcuts. To "convince" the biology of future generations to modify lifelong developmental trajectories—that is, to improve long-term offspring health via intergenerational interventions—we must strive to develop interventions that sustain, or at least mimic, longer timescale environmental change.

How might this goal be achieved? One possibility, which has gained some empirical support (see Kuzawa and Thayer 2011), is that intervening at several ages will have synergistic effects on offspring outcomes. For instance, the flow of nutrients and hormones across the placenta and in breast milk both appear to influence metabolism, growth, and long-term biological settings in offspring. Might supplementing the mother's diet during pregnancy *and* lactation, sending a signal of consistently improved conditions, have effects that are greater than the sum of their independent parts? Similarly, might the programming effects of favorable nutritional signals conveyed to breastfed infants via breast milk be enhanced if the mother received supplements during or prior to pregnancy, rather than during lactation alone? Recognizing that fetal growth is more directly contingent upon maternal metabolism than her dietary intake suggests yet other potential strategies. Although not without potential risks, directly manipulating maternal metabolism during gestation (for instance, by raising or lowering blood glucose) should have relatively direct and potentially large effects on fetal nutrient transfer and metabolic programming.

The notion of adaptive timescales underscores that interventions aimed at improving the gestational flow of resources across generations need to broaden the window of intervention to include the mother's own development, a point recently emphasized in the public health literature (Victora et al. 2008). Given evidence that early life nutrition has intergenerational effects, public health efforts to improve health and nutrition during early development, for instance, by supplementing nutritional intake or reducing the burden of common infections, can be understood as not only benefiting the infant but also the future offspring of that individual, who we may hypothesize will experience more favorable fetal nutrition and growth conditions as a result.

In a unique supplementation trial in Guatemala, young girls who were provided high-protein/calorie supplements as infants and young children, when compared with peers who received only a high calorie supplement, went on to give birth to larger babies decades later, pointing to the potential intergenerational power of a long-term developmental approach to nutritional intervention (Behrman et al. 2009).

The literature and ideas that we review here suggest that it is prudent to envision that the goal of our interventions is not only to alleviate stress in the present but also to find creative ways to mimic cues of sustained environmental improvement in the recent past (Kuzawa and Thayer 2011). By producing signals that indicate sustained nutritional improvements, we can hope that future generations of our long-lived species will alter developmental trajectories in ways that help limit the rising burden of chronic disease.

References

Addo OY, Stein AD, Fall CHD, Gigante DP, Guntupalli AM, Horta BL, Kuzawa CW, Lee N, Norris SA, Osmond C, Prabhakaran P, Richter LM, Sachdev HPS, Martorell R (2015) Parental childhood growth and offspring birthweight: pooled analyses from four birth cohorts in low and middle income countries. Am J Hum Biol 27(1):99–105

Armitage JA, Poston L, Taylor PD (2008) Developmental origins of obesity and metabolic syndrome: the role of maternal obesity. Front Horm Res 36:73–94

Barker D (1994) Mothers, babies, and disease in later life. BMJ, London

Barker DJ, Osmond C, Golding J, Kuh D, Wadsworth ME (1989) Growth in utero, blood pressure in childhood and adult life, and mortality from cardiovascular disease. Br Med J 298(6673):564–567

Bateson G (1963) The role of somatic change in evolution. Evolution 17(4):529

Bateson P (2001) Fetal experience and good adult design. Int J Epidemiol 30(5):928–934

Bateson P, Barker D, Clutton-Brock T, Deb D, D'Udine B, Foley RA, Gluckman P, Godfrey K, Kirkwood T, Lahr MM et al (2004) Developmental plasticity and human health. Nature 430(6998):419–421

Behrman JR, Calderon MC, Preston SH, Hoddinott J, Martorell R, Stein AD (2009) Nutritional supplementation in girls influences the growth of their children: prospective study in Guatemala. Am J Clin Nutr 90(5):1372–1379

Boiko J, Jaquet D, Chevenne D, Rigal O, Czernichow P, Levy-Marchal C (2005) In situ lipolytic regulation in subjects born small for gestational age. Int J Obes (Lond) 29(6):565–570

Ceesay SM, Prentice AM, Cole TJ, Foord F, Poskitt EM, Weaver LT, Whitehead RG (1997) Effects on birth weight and perinatal mortality of maternal dietary supplements in rural Gambia: 5 year randomised controlled trial. Br Med J 315(7111):786–790

Chung GC, Kuzawa CW (2014) Intergenerational effects of early life nutrition: maternal leg length predicts offspring placental weight and birth weight among women in rural Luzon, Philippines. Am J Hum Biol 26(5):652–659

Dörner G, Plagemann A, Rückert J, Götz F, Rohde W, Stahl F, Kürschner U, Gottschalk J, Mohnike A, Steindel E (1988) Teratogenetic maternofoetal transmission and prevention of diabetes susceptibility. Exp. Clin. Endocrinol 91(3):247–258

Drake AJ, Liu L (2010) Intergenerational transmission of programmed effects: public health consequences. Trends Endocrinol Metab 21(4):206–213

Franklin TB, Russig H, Weiss IC, Gräff J, Linder N, Michalon A, Vizi S, Mansuy IM (2010) Epigenetic transmission of the impact of early stress across generations. Biol Psychiatry 68(5):408–415

Frisancho AR (1993) Human adaptation and accommodation. University of Michigan Press, Ann Arbor

Frisancho AR, Guilding N, Tanner S (2001) Growth of leg length is reflected in socio-economic differences. Acta Med Auxol 33:47–50

Gluckman PD, Hanson M (2005) The fetal matrix: evolution, development, and disease. Cambridge University Press, New York

Guerrero-Bosagna C, Skinner MK (2012) Environmentally induced epigenetic transgenerational inheritance of phenotype and disease. Mol Cell Endrocrinol 354(1–2):3–8

Hancox RJ, Poulton R, Greene JM, McLachlan CR, Pearce MS, Sears MR (2009) Associations between birth weight, early childhood weight gain and adult lung function. Thorax 64:228–232

Harrison M, Langley-Evans SC (2009) Intergenerational programming of impaired nephrogenesis and hypertension in rats following maternal protein restriction during pregnancy. Br J Nutr 101:1020–1030

Heijmans BT, Tobi EW, Stein AD, Putter H, Blauw GJ, Susser ES, Slagboom PE, Lumey LH (2008) Persistent epigenetic differences associated with prenatal exposures to famine in humans. Proc Natl Acad Sci 105(44):17046–17049

Hucking K, Hamilton-Wessler M, Ellmerer M, Bergman RN (2003) Burst-like control of lipolysis by the sympathetic nervous system in vivo. J Clin Invest 111:257–264

Iliadou A, Cnattingius S, Lichtenstein P (2004) Low birthweight and Type 2 diabetes: a study on 11 162 Swedish twins. Int J Epidemiol 33(5):948–953

Kardjati S, Kusin JA, De With C (1988) Energy supplementation in the last trimester of pregnancy in East Java: I. Effect on birthweight. Br J Obstet Gynaecol 95:783–794

Kramer MS, Kakuma R (2003) Energy and protein intake in pregnancy. Cochrane Database Syst Rev 4, CD000032

Kuzawa CW (2005) Fetal origins of developmental plasticity: are fetal cues reliable predictors of future nutritional environments? Am J Hum Biol 17(1):5–21

Kuzawa CW (2007) Developmental origins of life history: growth, productivity, and reproduction. Am J Hum Biol 19(5):654–661

Kuzawa C (2008) The developmental origins of adult health: intergenerational inertia in adaptation and disease. In: Trevathan W, Smith E, McKenna J (eds) Evolutionary medicine and health: new perspectives. Oxford University Press, New York, pp 325–349

Kuzawa CW (2010) Beyond feast-famine: brain evolution, human life history, and the metabolic syndrome. In: Muehlenbein M (ed) Human evolutionary biology. Cambridge University Press, Cambridge

Kuzawa CW, Eisenberg DTAE (2012) Intergenerational predictors of birth weight in the Philippines: correlations with mother's and father's birth weight and a test of the maternal constraint hypothesis. PLoS One 7, e40905

Kuzawa CW, Eisenberg DT (2014) The long reach of history: intergenerational and transgenerational pathways to plasticity in human longevity. In: Weinstein M, Lane M (eds) Sociality, hierarchy, health: comparative biodemography: papers from a workshop. National Academies Press, Washington, DC, pp 65–94

Kuzawa CW, Pike IL (2005) Introduction. Fetal origins of developmental plasticity. Am J Hum Biol 17(1):1–4

Kuzawa CW, Sweet E (2009) Epigenetics and the embodiment of race: developmental origins of US racial disparities in cardiovascular health. Am J Hum Biol 21(1):2–15

Kuzawa C, Thayer Z (2011) Timescales of human adaptation: the role of epigenetic processes. Epigenomics 3(2):221–234

Kuzawa CW, Thayer Z (2012) Epigenetic Embodiment of Health and Disease: A Framework for Nutritional Intervention. Institute for Policy Research Northwestern University, Working Paper Series, WP-12–15

Lampl M, Kuzawa CW, Jeanty P (2002) Infants thinner at birth exhibit smaller kidneys for their size in late gestation in a sample of fetuses with appropriate growth. Am J Hum Biol 14(3):398–406

Lasker G (1969) Human biological adaptability: the ecological approach in physical anthropology. Science 166:1480–1486

Lawlor DA, Davey Smith G, Ebrahim S (2003) Association between leg length and offspring birthweight: partial explanation for the transgenerational association between birthweight and cardiovascular disease: findings from the British Women's Heart and Health Study. Paediatr Perinat Epidemiol 17:148–155

Levitt NS, Lambert EV, Woods D, Hales CN, Andrew R, Seckl JR (2000) Impaired glucose tolerance and elevated blood pressure in low birth weight, nonobese, young South African adults: early programming of cortisol axis. J Clin Endocrinol Metab 85(12):4611–4618

Martin RM, Davey Smith G, Frankel S, Gunnell D (2004) Parents' growth in childhood and the birth weight of their offspring. Epidemiology 15:308–316

Metzger BE, Lowe LP, Dyer AR, Trimble ER, Chaovarindr U, Coustan DR, Hadden DR, McCance DR, Hod M, McIntyre HD, Oats JJ, Persson B, Rogers MS, Sacks DA (2008) Hyperglycemia and adverse pregnancy outcomes. N Engl J Med 358(19):1991–2002

Mora JO, De Navarro L, Clement J, Wagner M, De Paredes B, Herrera MG (1978) The effect of nutritional supplementation on calorie and protein intake of pregnant women. Nutr Rep Int 17:217–228

Nwagwu MO, Cook A, Langley-Evans SC (2000) Evidence of progressive deterioration of renal function in rats exposed to a maternal low-protein diet in utero. Br. J. Nutr. 83(1):79–85

Pembrey ME, Bygren LO, KaatiG ES, Northstone K, Sjostrom M, Golding J (2006) Sex-specific, male-line transgenerational responses in humans. Eur J Hum Genet 14(2):159–166

Roseboom TJ, van der Meulen JH, Ravelli AC, Osmond C, Barker DJ, Bleker OP (2001) Effects of prenatal exposure to the Dutch famine on adult disease in later life: an overview. Mol Cell Endocrinol 185(1–2):93–98

Roseboom TJ, de Rooij S, Painter R (2006) The Dutch famine and its long-term consequences for adult health. Early Hum Dev 82:485–491

Rush D, Stein Z, Susser M (1980) A randomized controlled trial of prenatal nutritional supplementation in New York City. Pediatrics 65:683–697

Victora CG, Adair L, Fall C, Hallal PC, Martorell R, Richter L, Sachdev HS on behalf of the Maternal and Child Undernutrition Study Group (2008) Maternal and child undernutrition: consequences for adult health and human capital. Lancet 371:340–357

Wells JC (2003) The thrifty phenotype hypothesis: thrifty offspring or thrifty mother? J Theor Biol 221(1):143–161

West-Eberhard M (2003) Developmental plasticity and evolution. Oxford University Press, New York

Yliharsila H, Kajantie E, Osmond C, Forsen T, Barker DJ, Eriksson JG (2007) Birth size, adult body composition and muscle strength in later life. Int J Obes (London) 31(9):1392–1399

Chapter 3
Modeling Developmental Plasticity in Human Growth: Buffering the Past or Predicting the Future?

Jonathan C.K. Wells and Rufus A. Johnstone

Abstract Substantial variation in adult body size between human populations is widely assumed in part to represent adaptation to local ecological conditions. Developmental plasticity contributes to such variability; however, there is debate regarding how this early-life process can produce adaptation when environments change within the life span. We developed a simple mathematical simulation model, testing how human fetuses could tailor their growth to ecological conditions without being oversensitive and hence prone to extremes of growth. Data on Indian rainfall (1871–2004) were used as an index of ecological conditions. The simulation model allowed the comparison of different strategies for processing these time-series data regarding (a) the toleration of short-term ecological variability and (b) the prediction of conditions in adulthood. We showed that ecological information processing is favored in environments prone to long-term ecological trends. Once this strategy is adopted, resistance to short-term ecological perturbations can be achieved either by lengthening the duration of developmental plasticity or by accumulating multigenerational influences. A multigenerational strategy successfully dampens the transmission of the effects of ecological shocks to future generations, but it does not predict or enable offspring to respond to longer-term conditions. However, this strategy does allow fetal growth to be tailored to the likely supply of nutrition from the mother in the period after birth, during when extrinsic mortality risk is high. Our model has implications for public health policies aimed at addressing chronic malnutrition.

3.1 Introduction

The concept of adaptation assumes that organisms optimize their evolutionary fitness by improving their ability to survive and breed in a given environment. Adaptive variation in body size is predicted to emerge through cumulative

J.C.K. Wells (✉)
Childhood Nutrition Research Centre, UCL Institute of Child Health,
30 Guilford Street, London WC1N 1EH, UK
e-mail: jonathan.wells@ucl.ac.uk

R.A. Johnstone
Department of Zoology, University of Cambridge, Downing Street, Cambridge CB2 3EJ, UK

© Springer Science+Business Media New York 2017
G. Jasienska et al. (eds.), *The Arc of Life*, DOI 10.1007/978-1-4939-4038-7_3

trade-offs between the life history functions of survival, growth, and reproduction, as shaped by diverse ecological factors (Harvey et al. 1987; Hill 1993). In humans, variability in adult size has been associated with factors such as the thermal environment, energy supply, and mortality risk (Katzmarzyk and Leonard 1998; Walker et al. 2006). A portion of this variability appears to have occurred as a result of natural selection acting directly on genetic variability, since over 200 genes have now been associated with adult height in humans (Lango Allen et al. 2010), and the trait has high heritability (Silventoinen et al. 2003; though see Wells and Stock 2011).

In addition to genotype, phenotypic plasticity represents an alternative means whereby organisms can respond to ecological stresses (West-Eberhard 2003). It might appear intuitive that phenotypic plasticity should favor adaptation to ecological conditions—however, not all plasticity results in greater fitness (Via et al. 1995; Ellison and Jasienska 2007). For human body size, some of the variability in adult phenotype derives from growth variability early in the life course. Growth becomes increasingly canalized from early childhood on (Bogin 1999; Mei et al. 2004; Smith et al. 1976). This pattern, in which plasticity in many traits is greatest in early life, is widely prevalent across species (Bateson 2001; McCance 1962; Widdowson and McCance 1960). Such "developmental plasticity" is certainly one way in which phenotypic variability is shaped by ecological stresses. In humans, however, the primary period of plasticity occurs two decades prior to exposure to the adult environment, wherein many selective pressures relevant to reproductive fitness must act. This raises questions regarding the extent to which developmental plasticity in our species is indeed adaptive throughout the entire life course or under all ecological circumstances.

That human growth variability has an adaptive component over the *short* term is well established. In general, higher body weight at birth and during infancy is associated with greater survival early in life (Hogue et al. 1987; Kow et al. 1991; Victora et al. 2001), when extrinsic mortality risk is greatest. When nutritional supply is constrained, however, offspring grow slowly and appear to prioritize growth and development of some organs or body components at the cost of others (Hales and Barker 1992; Latini et al. 2004; Pomeroy et al. 2012).

The notion that early growth variability has *long*-term adaptive value is more controversial. If ecological stresses encountered in early life persist into later life, then developmental plasticity might promote fitness of the organism in its adult environment. For example, the thermal environment tends to be relatively consistent across broad global regions; hence, heat and cold stress early in life might induce beneficial adjustments in body size and proportions. Consistent with that hypothesis, ecogeographical analyses have demonstrated correlations of both adult phenotype and birth weight with heat stress (Roberts 1953; Katzmarzyk and Leonard 1998; Wells and Cole 2002; Wells 2012a), suggesting that human adaptation to the thermal environment begins in utero.

In volatile or unpredictable environments, however, deriving adult adaptation through early-life plasticity is inherently challenging. Paleoclimate evidence indicates that hominin evolution took place in increasingly stochastic environments (Bonnefille et al. 2004; Lisiecki and Raymo 2005; Potts 2012a, b; Trauth et al.

2005). Long-term "Milankovitch cycles" drive climate change over tens of thousands of years (Glantz 2001). Shorter-term climate cycles are also evident in the hominin paleoclimate record (Wang et al. 2008), including some that are analogous to contemporary El Niño-Southern Oscillation (ENSO) cycles (Hughen et al. 1999). These climate cycles, in turn, can be assumed to have introduced shorter-term ecological variability. Although phenotypic plasticity may potentially aid adaptation in such stochastic environments, developmental plasticity has a low degree of reversibility (Piersma and Drent 2003), and any adaptive benefits might not extend into adulthood.

This potential for "disconnect" between early-life plasticity and later-life adaptation is likely to have been exacerbated in recent human evolution by a substantial lengthening of the developmental period (Bogin and Smith 1996) which, paradoxically, may itself have been favored by stochastic environments (Wells 2012b). Whereas other female apes achieve reproductive maturity within a decade of birth (Galdikas and Wood 1990; Robson and Wood 2008), humans require around two decades to reach the same state. In volatile environments, this extended growth period decreases the likelihood that early plasticity will generate traits well suited to the ecological conditions encountered in adulthood. How then can developmental plasticity actually allow beneficial adaptation to the environment?

While ecological stresses may be evaluated in terms of direct, material environmental effects on growth, they may also be considered as a source of "information" about the quality of the environment with the potential to have more indirect, sustained effects (Bateson 2001). We can reframe the dilemma concerning timing of adaptive influences on developmental plasticity as follows: how might information received by organisms early in life be translated or processed in terms of adaptive growth patterns?

3.2   Developmental Plasticity as Information Processing

Several different models of developmental plasticity as information processing have been proposed previously. A general approach was offered by Bateson (2001); in this "weather forecast" model, the developing organism receives cues of impending environmental conditions and selects an appropriate developmental trajectory accordingly. Since physiological plasticity cannot be maintained indefinitely, a specific strategy must be selected early in life during a critical window of development. An accurate weather forecast is assumed to enable an appropriate future strategy, whereas an inaccurate forecast results in the organism being poorly prepared for its long-term environment. Key questions arising from this perspective are, first, what specific cues about the environment are obtained and, second, what broader ecological parameters do those cues index?

This "forecasting" framework has been extended by Gluckman and Hanson (2004a, b) in the form of the "predictive adaptive response" (PAR) hypothesis. This model assumes that developing offspring receive cues during pregnancy about the state of the environment and use them to predict the adult environment in which reproduction is likely to take place (Gluckman and Hanson 2004a, b; Gluckman et al. 2007). For example, offspring experiencing famine in early life are assumed to prepare for persisting famine in adulthood through enhancing traits such as insulin resistance and central adiposity. The challenge of this approach for a long-lived species like humans is that predictions about the environment must be accurate for several decades in the future and remain accurate, given that reproduction does not even begin until late in the second decade after birth.

This PAR hypothesis has been extensively criticized on several grounds, all related to the idea that such long-term forecasting is implausible (Jones 2005; Bogin et al. 2007; Rickard and Lummaa 2007; Wells 2007a, 2010, 2012c). First, spectral analyses of simulated or historically stochastic environments fail to support the hypothesis that current or recent-past conditions can predict future conditions (Wells 2007a; Baig et al. 2011). Second, mortality is highest in human foragers in early life, raising questions as to how "long-term anticipatory adaptation" could develop in traits already strongly exposed to selection earlier in the life course (Wells 2007a). Third, empirical data often contradict the predictions of the PAR hypothesis (Wells 2012c): for example, Gambians under seasonal energy scarcity do not develop insulin resistance following low birth weight (Moore et al. 2001). An alternative "silver spoon" hypothesis predicts that offspring receiving more early-life investment have higher reproductive fitness in all types of adult environments (Monaghan 2008). This hypothesis is supported in the comparative literature for a variety of vertebrate animal species, including humans (Monaghan 2008; Hayward et al. 2013).

An alternative "maternal capital hypothesis" emphasizes that the information processed by offspring during placental nutrition and lactation derives from the maternal phenotype, rather than directly from the external environment (Wells 2003, 2010, 2012c). Notably, human birth weight is only moderately depressed during maternal famine and only moderately increased following maternal supplementation, indicating that maternal physiology buffers the fetus from short-term fluctuations (Wells 2003). Exposure of the fetus to maternal phenotype, representing the cumulative effect of the nutritional environment experienced during development (Emanuel et al. 2004; Hyppönen et al. 2004; Jasienska 2009), as well as any previous reproductive experience for the mother, means that "short-term fluctuations [are] smoothed out to provide a more reliable rating of environmental quality" (Wells 2003). In this approach, adaptation through developmental plasticity is considered to be not to long-term future conditions, but to "maternal capital" (Wells 2003, 2010, 2012c). This approach also emphasizes that offspring plasticity makes possible "maternal effects" that benefit maternal as well as offspring fitness (Wells 2003, 2007b).

Elements of both the PAR hypothesis and maternal buffering have been presented by Kuzawa (2005) in his model of "intergenerational inertia." As with the maternal capital hypothesis, Kuzawa argued that maternal phenotype buffers the offspring from short-term ecological perturbations and provides a smoothed signal of ecological conditions deriving from matrilineal experience. As in the PAR hypothesis, however, Kuzawa assumes that this smoothed signal early in life aids the long-term prediction of ecological conditions, by providing "a 'best guess' of conditions likely to be experienced in the future" (Kuzawa and Bragg 2012).

These three models, therefore, while they all treat developmental plasticity as adaptive, have significant differences (Box 1). One difference involves the timescale

---

Box 1: Different Models of the Adaptive Nature of Developmental Plasticity

Hypothesis	Authors and reference	Source of ecological cues	Target of adaptation
Maternal capital	Wells (2003, 2010)	Maternal phenotype, integrating effects of maternal development (and hence grandmaternal phenotype), and life history strategy	Matching early offspring growth to maternal nutritional resources
Predictive adaptive response	Gluckman and Hanson (2004a, b)	External environment	Matching adult phenotype to adult reproductive environment
Intergenerational inertia	Kuzawa (2005)	Maternal and matrilineal phenotype	Matching adult phenotype to adult reproductive environment

We aimed to develop a simple mathematical model that enables the evaluation of different strategies by which offspring can obtain information early in life on environmental conditions relevant to fitness. We addressed three different types of ecological variability, as depicted schematically in Fig. 3.1. First, the environment may be subject to clear annual cycles, with peaks and troughs in ecological productivity. Second, the environment may be subject to systematic trends, such that ecological productivity may rise or fall over lengthy time periods, as might occur through larger climate trends. Third, the environment may be subject to irregular, unpredictable "extreme events," which superimpose major perturbations on other, more consistent patterns. Each of these three types of variability can be detected in the segment of the climate record relevant to human evolution, as well as in recent decades. For example, India experiences annual climate cycles, local systematic trends in rainfall, and irregular ENSO events that provoke monsoon failure (Glantz 2001; Guhathakurta and Rajeevan 2007).

To operationalize this approach, we used rainfall data from India as an index of ecological productivity and considered the kind of information that could be extracted from this record and processed adaptively by humans early in life, during the period of greatest developmental plasticity. We first considered the conditions under which it pays offspring to process information at all; then, having demonstrated that information processing can indeed be adaptive, we considered how different kinds of ecological variability can be adaptively translated and processed.

(continued)

Box 1 (continued)

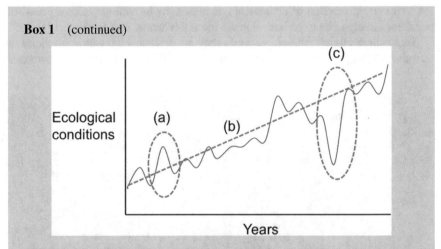

Fig. 3.1 A schematic diagram of variation in ecological conditions over time, illustrating three components of variability: (**a**) regular cycles, such as seasonality, (**b**) long-term systematic trends such as climate change, and (**c**) extreme perturbations, such as El Niño-Southern Oscillation (ENSO) events

over which the information is acquired. Another difference concerns the stage of the life course at which the adaptive response is assumed to be targeted. To date, debate over these contrasting approaches has been conducted through verbal arguments, with little systematic testing of competing hypotheses.

3.3 The Rationale of the Model

Data on monthly rainfall (R) for the period January 1871–December 2002 were obtained from the IRI/LDEO Climate Data Library (http://iridl.ldeo.columbia. edu), for the region of India designated "core monsoon," located at 76°E, 22.5°N. The typical pattern of variability is shown in Fig. 3.2, illustrating both annual variability and the 1900 ENSO event. We treated rainfall level as a proxy for food availability, an ecological cue especially relevant to the organism during development.

The primary outcome variable was birth weight (B), and our model assumed that this trait is subject to two opposing tensions. On the one hand, below a certain threshold, birth weight is inadequate for sustaining an infant. This favors increasing birth weight. On the other hand, above a certain upper threshold, fetal growth exceeds the mother's capacity to extract energy from current or recent environments. In other words, we considered that the challenge for the offspring is to calibrate its fetal growth to ecological conditions in a way that avoids these extreme phenotypes.

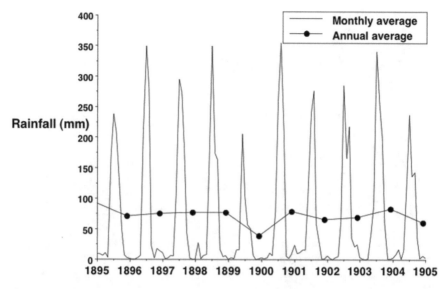

Fig. 3.2 The pattern of rainfall in the Indian data set used for this model, showing raw monthly values and average annual values

Birth weight correlates strongly with adult size (Li et al. 2003; Sayer et al. 2004; Euser et al. 2005). While larger birth weight brings higher fitness returns in adult life (Wells 2007b), the trajectory of growth toward large adult size is inherently constrained by the availability of maternal resources during early life. This assumption is supported by evidence that human gestation length is constrained by energetic rather than biomechanical factors (Leutenegger 1982; Ellison 2008; Dunsworth et al. 2012).

We posed two key questions: how sensitive should birth weight be to environmental cues, and how can useful information be extracted by the offspring from the crude ecological record?

3.4   The Basic Model

The model was constructed using a succession of algorithms to simulate how human offspring might process crude ecological information to derive an adaptive growth strategy (Box 2). In order to express our results in a way that facilitates comparison with empirical birth weight data, we assigned an average value for B of 3 kg (in other words, we treat this value as a population average that is present now because it maximized fitness in the recent past). The average value for R over the entire period was 79.7 mm; hence, we converted all R-data entered into the model into output B-data by dividing by (79.7/3), or 26.6.

Box 2: Algorithms Tested Using the Model

The basic model	Matching birth weight to rainfall in the last month
Simple smoothing	Extracting a smoothed signal from rainfall using a rolling average
Lengthening pregnancy	Increasing the time period over which the rolled average is collected
Minimal information processing	Weighting birth weight toward a fixed value and reducing the contribution of the rolling average
Generational effects	Collecting rolled averages from the plastic periods of two or more successive generations

Variability in B can be crudely assessed using the coefficient of variation (CV). Although a large CV provides one indication of the risk of excessive or insufficient birth weight values, it is also helpful to see how this risk is distributed. We therefore calculated "fitness penalties," as the difference between the actual B value and 3 kg. The larger the difference, the more birth weight was inadequate or excessive. These fitness values could be expressed in absolute values or squared to make positive and negative penalties equivalent.

Using this model, we considered a range of information-processing algorithms whereby growth could be calibrated to environmental conditions so as to reduce fitness penalties.

3.5   Simple Smoothing

The crudest strategy for adjusting reproductive output to ecological conditions would be for the organism to track R on a month-by-month basis, allowing B to respond to the last available signal of R prior to birth, such that B is proportional to R:

$$B \propto R \qquad (3.1)$$

In the simulation, the standard deviation of R was 109.3 mm. As R fluctuated substantially across every annual cycle [represented by the coefficient of variation (CV) of 137.2 %] and B varied proportionately, direct tracking of R produced minimum and maximum values for B of 0 and 18.9 kg, respectively, generating a very high proportion of nonviable B values and hence high fitness penalties.

A simple way to model dampened sensitivity to crude variability in ecological conditions would be to average the response to R over a longer time period (Wells 2003; Kuzawa 2005). For example, the duration of pregnancy allows ecological information to act on the phenotype directly over a 9-month period, which can be represented using a rolling average, as shown here:

$$B \propto 1/9\left(\sum_{t-9}^{t} R\right) \tag{3.2}$$

This rolling average still varied between successive 9-month periods, and it itself had a CV of 40.3 %. In order to test whether the variability of B could be further reduced by extending the duration of the period of information processing (i.e., effectively increasing the duration of pregnancy), rolling averages were also calculated over 18-, 27-, 36-, and 45-month periods. In each case, the CV for birth weight was calculated, along with squared fitness penalties.

Figure 3.3 illustrates the decline in CV and the increase in minimum birth weight in relation to lengthening the duration of pregnancy. Integrating information over 36 months or more reduced the CV of birth weight to ~10 % and the range to 2–3.7 kg. The fitness penalty declined substantially up to 27 months, but little thereafter. Thus, one way to minimize the likelihood of unviable birth weight would be to "drip-feed" ecological information into the offspring phenotype over a lengthy time period. However, lengthening pregnancy clearly has other penalties in terms of maternal fitness, as it reduces fertility rate over the entire reproductive career.

Paradoxically, a similar stability in birth weight could be obtained simply by dampening sensitivity to ecological conditions—in other words, by minimizing the weight given to current ecological information. This is equivalent to increasing a genetic constraint on birth weight and reducing the magnitude of plasticity. For example, a simple strategy in which birth weight was a function of a relatively fixed expectation (3 kg), and allowed to vary only modestly around this value (say, 33 % of the actual difference between the 9-month average and 3 kg), also produced very low fitness penalties. When this strategy was used, the CV of birth weight obtained was 12 %, the minimum birth weight was 2.19 kg, and the average fitness penalty was 0.13. If fitness penalties can be minimized without significant

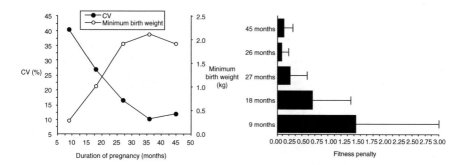

Fig. 3.3 The effect of increasing the period of information processing (equivalent to increasing the duration of pregnancy) on variability in birth weight. (**a**) Lengthening pregnancy decreases the coefficient of variation and increases the minimum birth weight value. (**b**) Lengthening pregnancy decreases the fitness penalties arising from suboptimal birth weights

information processing, then there must be another axis of ecological variability, not addressed in the modeling so far, that is important.

3.6   Long-Term Ecological Trends

The limitations of the fixed "minimal processing" strategy become evident if we manipulate the raw rainfall data to introduce a downward secular trend in over time, thus simulating a long-term ecological trend that was not actually evident in our 130-year rainfall data.

The distribution of raw error in rainfall over time was plotted for the two scenarios: (a) no trend over time versus (b) a downward trend. The resulting CV of birth weight was 12.1 % for no trend and 13.8 % for the downward trend. The fitness penalty was 0.13 (SD 0.14) for no trend, but 0.17 (SD 0.19) for the downward trend, indicating relatively similar overall fitness in both scenarios. However, as illustrated in Fig. 3.4, the distribution of the raw (unsquared) fitness penalties for each scenario showed that whereas penalties were randomly distributed in the absence of any trend in rainfall, they changed systematically across time with the downward trend.

The "minimal processing" strategy systematically produced birth weight values lower than the optimum during the early part of this downward trend and higher values than the optimum during the end of the trend. This means that should a trend ultimately lead to the environment stabilizing at a lower level of rainfall, a "minimal processing" strategy would remain locked into higher fitness penalties, with birth weights exceeding adaptive levels for the available energy supply. Thus, a fixed strategy cannot accommodate long-term trends and would be outcompeted by the plastic smoothing strategy described above, allowing sensitivity to ecological signals.

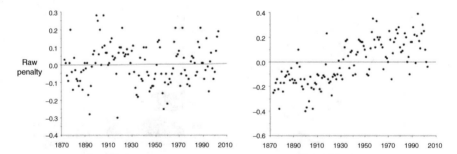

Fig. 3.4 The distribution of fitness penalties when information processing is minimal in different types of environment. (**a**) In a stable environment, fitness penalties are randomly distributed. (**b**) In an environment systematically declining in productivity over time, fitness penalties are not randomly distributed and may become systematically high since the phenotype cannot adjust to the ecological change

3.7   Generational Effects

On the basis that information processing is indeed valued, but cannot be obtained through extending the duration of pregnancy, we considered an alternative algorithm that could process greater quantities of ecological data.

Since plasticity in each generation is greatest in early life, a maternal effect was introduced to simulate conditions during the mother's own early life. Assuming a generation time of 20 years, the maternal effect was operationalized by integrating rolling averages from both the current 9-month period (G0) and a second 9-month period 20 years earlier (G1). In this model, therefore, both maternal developmental conditions and current maternal conditions could shape the offspring B value. This represents a one-generation maternal effect, i.e.,

$$B \propto 1/18 \left\{ \left(\sum_{t-9}^{t} R \right) + \left(\sum_{t-249}^{t-240} R \right) \right\} \tag{3.3}$$

The model was further developed by adding simulations (G1 through G4) that included up to five generations of additional maternal effects. The final model, therefore, simulates the integration by the offspring of information on the current 9-month period (G0), as well as 9-month periods 20 years ago (maternal effects, G1), 40 years ago (grandmaternal effects, G2), 60 years ago (great-grandmaternal effects, G3), 80 years ago (great-great-grandmaternal effects, G4), and 100 years ago (great-great-great-grandmaternal effects, G5):

$$B \propto 1/54 \left\{ \left(\sum_{t-9}^{t} R \right) + \left(\sum_{t-249}^{t-240} R \right) + \left(\sum_{t-489}^{t-480} R \right) + \left(\sum_{t-729}^{t-720} R \right) + \left(\sum_{t-969}^{t-960} R \right) + \left(\sum_{t-1209}^{t-1200} R \right) \right\} \tag{3.4}$$

Each of these models was run using R-data from the period of 1970–2002, and the range and CV of B were calculated. As shown in Fig. 3.5, there was a decline in the CV of B with each additional generation of lag for the first three generations; adding in further generational effects produced a negligible difference in the outcome.

Since these findings indicated no substantial benefits of integrating data across more than three generations preceding the focal developmental period, all subsequent models used just one or three generations of lag—that is, incorporating either maternal effects or great-grandmaternal effects.

3.8   Buffering Extreme Events

We next tested the effect of a sudden drastic decline in R in the past, as in an ENSO event. The average annual value for R in the entire data set was 79.7 mm; however, 3 years during this period reflected particularly severe ENSO events, during which the annual average rainfall was markedly lower: 1899 (38.0 mm), 1920 (43.8 mm), and 1970 (50.9 mm).

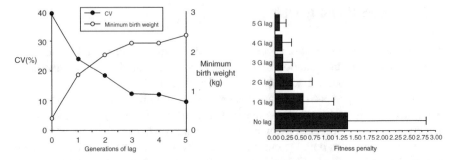

Fig. 3.5 The effect of introducing maternal effects (which buffer short-term ecological variability) across increasing numbers of generations on variability in birth weight. (**a**) For the first three generations, each extra generation of lag increases buffering, decreases the coefficient of variation, and increases the minimum value of birth weight. (**b**) For the first three generations, increased buffering also decreases the fitness penalties arising from suboptimal birth weights. However, for all outcomes, benefits are minimal when extending maternal effects across four or five generations

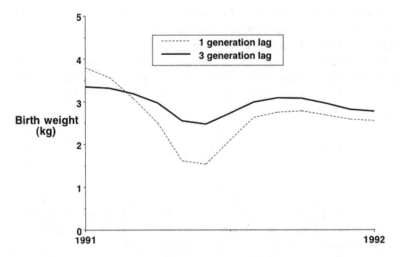

Fig. 3.6 The effect of one versus three generations of maternal effects, or buffering ecological variability, on offspring birth weight following an extreme event at the time of the mother's birth. The impact of this maternal stress on the offspring's birth weight is greatly reduced by the multigenerational buffering process

The model assumed that offspring were born in 1990, 20 years after the 1970 ENSO event, which had therefore occurred during the mothers' own fetal life. Birth weight patterns for offspring were calculated incorporating either one or three generations of maternal buffering of intergenerational environmental variation, as explained above. When maternal effects only were incorporated, birth weight of offspring dropped as low as ~1.5 kg, whereas when great-grandmaternal effects were used, the lowest birth weight was ~2.5 kg (Fig. 3.6).

This simulation therefore showed that multigenerational damping of environmental effects has the potential to be very successful at buffering offspring from the consequences of extreme short-term perturbations experienced by their mother during her own development. Only a modest adverse effect is propagated to future generations. In the event that several additional generations are spared such extreme events during fetal life, this adverse effect would be expected to wash out entirely.

3.9 Predicting the Future

The final model tested the capacity of offspring birth weight to predict subsequent ecological conditions, as simulated using two different strategies. In the first approach, birth weight based on a 9-month rolling average was associated with rainfall in the following month ($r=0.14$, $p<0.0001$) and with maternal phenotype at 1 year ($r=0.79$, $p<0.0001$), but not with rainfall 20 years in the future, either averaged over 1 year ($r=0.03$, ns) or 20 years ($r=0.05$, ns).

Using the three-generation lagged model for birth weight, there was no correlation between offspring phenotype at birth and rainfall in the subsequent month ($r=-0.09$, ns) (Fig. 3.7a). On the other hand, a strong correlation was found between offspring phenotype at birth and maternal phenotype 1 year later ($r=0.62$, $p<0.0001$) (Fig. 3.7b). Since offspring phenotype is derived from the maternally smoothed signal, this demonstrates how maternal phenotype can represent a stable ecological signal in the immediate postnatal period, regardless of how the external environment actually changes during this time. Figure 3.7c, d shows, further, that there was no correlation between offspring phenotype at the time of birth and a 1-year average of R 20 years in the future (i.e., ecological conditions when the offspring will reach reproductive maturity) ($r=-0.08$, ns) or a 10-year average of R commencing 20 years in the future (i.e., ecological conditions during the offspring's reproductive career) ($r=-0.08$, ns). Thus, the results of this simulation suggest that information extracted from maternal phenotype in early life cannot match the offspring's phenotype with ecological conditions encountered in adult life.

3.10 Discussion

We used a simple mathematical simulation model based on an actual historical data set of ecological conditions to investigate different strategies whereby offspring might adaptively process information relevant to fitness during early, sensitive periods of development. The challenge facing the offspring was to be able to respond to ecological stresses in early life, but not to the extent that growth patterns became extreme. We then considered whether strategies that solved this dilemma could match the organism's phenotype with ecological conditions in adulthood. The main findings were as follows.

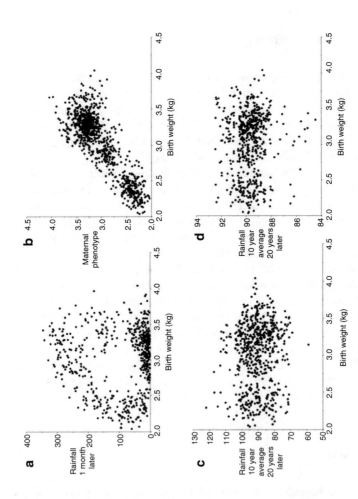

Fig. 3.7 Associations between offspring birth weight (derived from three generations of maternal buffering) and four different sets of subsequent ecological conditions. The results show (**a**) no significant correlation between birth weight and rainfall in the month immediately following birth, (**b**) a strong correlation between birth weight and maternal phenotype 1 year later ($r=0.62$, $p<0.0001$), (**c**) no correlation between maternal phenotype at the time of birth and for a 12-month average commencing 20 years in the future (representing the time of attaining reproductive maturity), (**d**) no correlation between maternal phenotype at the time of birth and a 10-year average commencing 20 years in the future (representing the duration of the reproductive career)

First, even crude processing of available information is a more adaptive strategy than ignoring ecological cues altogether and has the potential to reduce the likelihood of unviable birth weight while also tracking long-term ecological trends. Although a new genetic set point for birth weight could evolve over many generations, such an evolutionary process would be very slow relative to the actual rate of ecological change. Consistent with this argument, relatively little of the variance in human birth weight can be accounted for by genetic factors (Magnus et al. 2001; Lunde et al. 2007).

Second, while lengthening the period over which ecological information is collected should help buffer short-term variability, there is a limit to how far this strategy can be applied, given that it imposes costs on maternal fertility. Thus, lengthier pregnancies eventually become an inefficient way of smoothing ecological signals and constraining extreme growth patterns. This result is supported by recent work demonstrating that the maternal energy budget is insufficient to support lengthier human pregnancies (Ellison 2008; Dunsworth et al. 2012).

Third, an alternative, beneficial damping effect can be introduced by incorporating maternal effects, and this effect can be multiplied across several successive generations. In our simulation, little additional benefit accrued beyond three generations of lag; however, the outcomes of this scenario might vary according to the degree of ecological stochasticity. Biological mechanisms for such lags may involve epigenetic marks, though the available evidence suggests that these are only rarely transmitted directly across generations (Youngson and Whitelaw 2008; Hackett et al. 2013). In the female line, the ovum that contributes to each offspring has already been exposed to the maternal uterine environment—and hence the grandmaternal phenotype—allowing grand maternal nutritional effects to be transmitted directly to grand offspring (Youngson and Whitelaw 2008; Drake and Liu 2010).

Fourth, such a multigenerational damping strategy could be very effective at avoiding the propagation of effects of severe ecological shocks to future generations. A mother born during an ENSO event is relatively well buffered by the three-generation lag, so that her own offspring is only moderately below the optimal birth weight. We consider this protective effect especially valuable, since size in early life is highly predictive of infant survival (Hogue et al. 1987), and infancy is the period of greatest extrinsic mortality risk in our species (Kelly 1995).

Fifth, although the ecological information extracted via the maternal phenotype allowed accurate short-term predictions using that maternal source of information, none of the strategies for information processing using in our simulations demonstrated any ability to predict long-term future conditions. Neither a short-term index, as proposed by (Gluckman and Hanson 2004a, b; Gluckman et al. 2007), nor an index smoothed across several ancestral generations, as proposed by Kuzawa (2005), showed any correlation with long-term future conditions. There was, however, a short-term correlation between offspring birth phenotype and maternal phenotype 1 year after birth.

These simulations therefore provide support for the maternal capital model of developmental plasticity (Wells 2003, 2010, 2012c), but not for models that assume that developmental plasticity enables fetal/infant phenotype to vary in anticipation of future ecological conditions (Kuzawa 2005; Gluckman and Hanson 2004a, b). The results of these simulations are also consistent with epidemiological studies which have shown that fetal growth is relatively resistant to short-term spikes or troughs in maternal energy intake, as reviewed previously (Wells 2003; Kuzawa 2005). The results regarding the inability to predict future conditions, moreover, are consistent with previous simulations (Wells 2007a; Baig et al. 2011).

The fact that short-term predictions are possible in our model shows that the off-spring can make a prediction about the nutritional supply likely to be available immediately after birth, during the lactation period. Since both the signals being processed and the nutritional supply after birth derive from maternal phenotype, this effectively means that the fetus can calibrate its growth to the likely supply of breast milk during infancy.

This is a specific prediction of the maternal capital hypothesis which, unlike other models, assumes that the primary fitness benefit of developmental plasticity is to enable a close match between offspring growth trajectory and maternal phenotype (Wells 2003; 2010; 2012c). However, another key reason why selection favors a match between maternal phenotype and offspring growth is that every fetus must avoid growing beyond the dimensions of the maternal pelvis to avoid the risk of cephalon-pelvic disproportion (Wells et al. 2012; Wells 2015).

The use of a simulation model to explore human adaptation has some strengths, as it allows comparison of different strategies across several generations, an approach rarely possible using data on humans themselves. However, the model also neglected some potentially important issues. For example, information does not only enter phenotype during the 9-month period of pregnancy each generation. A more realistic model would allow information to enter phenotype over longer time periods; hence, our model only investigates the effect of multigenerational lags acting on pregnancy. Similarly, human developmental plasticity extends into infancy, and growth during infancy can resolve some of the variability that characterizes birth weight; however, this was not addressed.

Despite these limitations, the model is valuable in demonstrating how maternal effects can damp ecological perturbations including extreme events. The same buffering process may have been important in enabling humans to migrate between contrasting ecological niches, by damping offspring from sudden shifts in nutritional supply during the most sensitive period of development (Wells 2012c). Thus, the fact that developmental plasticity does not allow forecasting of long-term adult environments does not mean that the process has no adaptive value. Rather, developmental plasticity in early life appears of greatest value in promoting survival during early life, a time when extrinsic mortality risk is high. This issue has important implications for public health policies aimed at addressing chronic malnutrition, as our findings suggests that continued maternal buffering via lactation may benefit prevent catch-up "overshoot" in the early postnatal period.

References

Baig U, Belsare P, Watve M et al (2011) Can thrifty gene(s) or predictive fetal programming for thriftiness lead to obesity? J Obes 861049

Bateson P (2001) Fetal experience and good adult design. Int J Epidemiol 30:928–934

Bogin B (1999) Patterns of human growth, 2nd edn. Cambridge University Press, Cambridge

Bogin B, Smith BH (1996) Evolution of the human life cycle. Am J Hum Biol 8:703–716

Bogin B, Silva MI, Rios L (2007) Life history trade-offs in human growth: adaptation or pathology? Am J Hum Biol 19:631–642

Bonnefille R, Potts R, Chalié F et al (2004) High-resolution vegetation and climate change associated with Pliocene Australopithecus afarensis. Proc Natl Acad Sci U S A 101:12125–12129

Drake AJ, Liu L (2010) Intergenerational transmission of programmed effects: public health consequences. Trends Endocrinol Metab 21:206–213

Dunsworth HM, Warrener AG, Deacon T et al (2012) Metabolic hypothesis for human altriciality. Proc Natl Acad Sci U S A 109:15212–15216

Ellison PT (2008) Energetics, reproductive ecology, and human evolution. PaleoAnthropology 2008:172–200

Ellison PT, Jasienska G (2007) Constraint, pathology, and adaptation: how can we tell them apart? Am J Hum Biol 19:622–630

Emanuel I, Kimpo C, Moceri V (2004) The association of maternal growth and socio-economic measures with infant birthweight in four ethnic groups. Int J Epidemiol 33:1236–1242

Euser AM, Finken MJ, Keijzer-Veen MG et al (2005) Associations between prenatal and infancy weight gain and BMI, fat mass, and fat distribution in young adulthood: a prospective cohort study in males and females born very preterm. Am J Clin Nutr 81:480–487

Galdikas BM, Wood JW (1990) Birth spacing patterns in humans and apes. Am J Phys Anthropol 83:185–191

Glantz M (2001) Currents of change: impacts of El Nino and La Nina on climate and society, 2nd edn. Cambridge University Press, Cambridge

Gluckman P, Hanson M (2004a) The fetal matrix: evolution, development and disease. Cambridge University Press, Cambridge

Gluckman PD, Hanson MA (2004b) The developmental origins of the metabolic syndrome. Trends Endocrinol Metab 15:183–187

Gluckman PD, Hanson MA, Beedle AS (2007) Early life events and their consequences for later disease: a life history and evolutionary perspective. Am J Hum Biol 19:1–19

Guhathakurta P, Rajeevan M (2007) Trends in the rainfall pattern over India. Int J Climatol 28:1453–1469

Hackett JA, Sengupta R, Zylicz JJ (2013) Germline DNA demethylation dynamics and imprint erasure through 5-hydroxymethylcytosine. Science 339:448–452

Hales CN, Barker DJ (1992) Type 2 (non-insulin-dependent) diabetes mellitus: the thrifty phenotype hypothesis. Diabetologia 35:595–601

Harvey PH, Martin RD, Clutton-Brock TH (1987) Life histories in comparative perspective. In: Smuts BB, Cheney DL, Seyfarth RM et al (eds) Primate societies. University of Chicago Press, Chicago, pp 181–196

Hayward AD, Rickard IJ, Lummaa V (2013) Influence of early-life nutrition on mortality and reproductive success during a subsequent famine in a preindustrial population. Proc Natl Acad Sci U S A 110:13886–13891

Hill K (1993) Life history theory and evolutionary anthropology. Evol Anthropol 2:78–88

Hogue CJ, Buehler JW, Strauss LT et al (1987) Overview of the National Infant Mortality Surveillance (NIMS) project—design, methods, results. Public Health Rep 102:126–138

Hughen KA, Schrag DP, Jacobsen SB (1999) El Nino during the last interglacial period recorded by a fossil coral from Indonesia. Geophys Res Let 26:3129–3132

Hyppönen E, Power C, Smith GD (2004) Parental growth at different life stages and offspring birthweight: an intergenerational cohort study. Paediatr Perinat Epidemiol 18:168–177

Jasienska G (2009) Low birth weight of contemporary African Americans: an intergenerational effect of slavery? Am J Hum Biol 21:16–24

Jones JH (2005) Fetal programming: adaptive life-history tactics or making the best of a bad start? Am J Hum Biol 17:22–33

Katzmarzyk PT, Leonard WR (1998) Climatic influences on human body size and proportions: ecological adaptations and secular trends. Am J Phys Anthropol 106:483–503

Kelly RL (1995) The foraging spectrum. Smithsonian Institution Press, Washington, DC

Kow F, Geissler C, Balasubramaniam E (1991) Are international anthropometric standards appropriate for developing countries? J Trop Pediatr 37:37–44

Kuzawa CW (2005) Fetal origins of developmental plasticity: are fetal cues reliable predictors of future nutritional environments? Am J Hum Biol 17:5–21

Kuzawa CW, Bragg JM (2012) Plasticity in human life history strategy: implications for contemporary human variation and the evolution of genus Homo. Curr Anthropol 52(Suppl 6):S369–S382

Lango Allen H, Estrada K, Lettre G et al (2010) Hundreds of variants clustered in genomic loci and biological pathways affect human height. Nature 467:832–838

Latini G, De Mitri B, Del Vecchio A (2004) Foetal growth of kidneys, liver and spleen in intrauterine growth restriction: "programming" causing "metabolic syndrome" in adult age. Acta Paediatr 93:1635–1639

Leutenegger W (1982) Encephalization and obstetrics in primates with particular reference to human evolution. In: Armstrong E, Falk D (eds) Primate brain evolution: methods and concepts. Plenum Press, New York, pp 85–95

Li H, Stein AD, Barnhart HX et al (2003) Associations between prenatal and postnatal growth and adult body size and composition. Am J Clin Nutr 77:1498–1505

Lisiecki LE, Raymo ME (2005) A plio-pleistocene stack of 57 globally distributed benthic $\delta^{18}O$ records. Paleoceanography 20

Lunde A, Melve KK, Gjessing HK et al (2007) Genetic and environmental influences on birth weight, birth length, head circumference, and gestational age by use of population-based parent-offspring data. Am J Epidemiol 165:734–741

Magnus P, Gjessing HK, Skrondal A et al (2001) Paternal contribution to birth weight. J Epidemiol Community Health 55:873–877

McCance RA (1962) Food, growth, and time. Lancet 2:621–626, 671–676

Mei Z, Grummer-Strawn LM, Thompson D et al (2004) Shifts in percentiles of growth during early childhood: analysis of longitudinal data from the California Child Health and Development Study. Pediatrics 113:e617–e627

Monaghan P (2008) Early growth conditions, phenotypic development and environmental change. Philos Trans R Soc B 363:1635–1645

Moore SE, Halsall I, Howarth D (2001) Glucose, insulin and lipid metabolism in rural Gambians exposed to early malnutrition. Diabet Med 18:646–653

Piersma T, Drent J (2003) Phenotypic flexibility and the evolution of organismal design. Trends Ecol Evol 18:228–233

Pomeroy E, Stock JT, Stanojevic S et al (2012) Trade-offs in relative limb length among Peruvian children: extending the thrifty phenotype hypothesis to limb proportions. PLoS One 7, e51795

Potts R (2012a) Evolution and environmental change in early human prehistory. Annu Rev Anthropol 41:151–167

Potts R (2012b) Environmental and behavioral evidence pertaining to the evolution of early Homo. Curr Anthropol 53(Suppl 6):S299–S317

Rickard IJ, Lummaa V (2007) The predictive adaptive response and metabolic syndrome: challenges for the hypothesis. Trends Endocrinol Metab 18:94–99

Roberts DF (1953) Body weight, race and climate. Am J Phys Anthropol 11:533–558

Robson SL, Wood B (2008) Hominin life history: reconstruction and evolution. J Anat 212:394–425

Sayer AA, Syddall HE, Dennison EM et al (2004) Birth weight, weight at 1y of age, and body composition in older men: findings from the Hertfordshire Cohort Study. Am J Clin Nutr 80:199–203

Silventoinen K, Sammalisto S, Perola M et al (2003) Heritability of adult body height: a comparative study of twin cohorts in eight countries. Twin Res 6:399–408

Smith DW, Truog W, Rogers JE et al (1976) Shifting linear growth during infancy: illustration of genetic factors in growth from fetal life through infancy. J Pediatr 89:225–230

Trauth MH, Maslin MA, Deino A et al (2005) Late Cenozoic moisture history of East Africa. Science 309:2051–2053

Via S, Gomulkiewicz R, De Jong G et al (1995) Adaptive phenotypic plasticity: consensus and controversy. Trends Ecol Evol 10:212–217

Victora CG, Barros FC, Horta BL et al (2001) Short-term benefits of catch-up growth for small-for-gestational-age infants. Int J Epidemiol 30:1325–1330

Walker R, Gurven M, Hill K et al (2006) Growth rates and life histories in twenty-two small-scale societies. Am J Hum Biol 18:295–311

Wang Y, Cheng H, Edwards RL et al (2008) Millennial- and orbital-scale changes in the East Asian monsoon over the past 224,000 years. Nature 451:1090–1093

Wells JC (2003) The thrifty phenotype hypothesis: thrifty offspring or thrifty mother? J Theor Biol 221:143–161

Wells JC (2007a) Flaws in the theory of predictive adaptive responses. Trends Endocrinol Metab 18:331–337

Wells JC (2007b) The thrifty phenotype as an adaptive maternal effect. Biol Rev 82:143–172

Wells JC (2010) Maternal capital and the metabolic ghetto: an evolutionary perspective on the transgenerational basis of health inequalities. Am J Hum Biol 22:1–17

Wells JC (2012a) Ecogeographical associations between climate and human body composition: analyses based on anthropometry and skinfolds. Am J Phys Anthropol 147:169–186

Wells JC (2012b) Ecological volatility and human evolution: a novel perspective on life history and reproductive strategy. Evol Anthropol 21:277–288

Wells JC (2012c) A critical appraisal of the predictive adaptive response hypothesis. Int J Epidemiol 41(1):229–235

Wells JC (2015) Between Scylla and Charybdis: renegotiating resolution of the 'obstetric dilemma' in response to ecological change. Philos Trans R Soc Lond B Biol Sci 370:20140067

Wells JC, Cole TJ (2002) Birth weight and environmental heat load: a between-population analysis. Am J Phys Anthropol 119:276–282

Wells JC, Stock JT (2011) Re-examining heritability: genetics, life history and plasticity. Trends Endocrinol Metab 22:421–428

Wells JC, DeSilva JM, Stock JT (2012) The obstetric dilemma: an ancient game of Russian roulette, or a variable dilemma sensitive to ecology? Am J Phys Anthropol 149(Suppl 55):40–71

West-Eberhard MJ (2003) Developmental plasticity and evolution. Oxford University Press, Oxford

Widdowson EM, McCance RA (1960) Some effects of accelerating growth. I. General somatic development. Proc R Soc Lond B Biol Sci 152:188–206

Youngson NA, Whitelaw E (2008) Transgenerational epigenetic effects. Annu Rev Genomics Hum Genet 9:233–257

Chapter 4
Development of Human Sociosexual Behavior

Peter B. Gray and Matthew H. McIntyre

Abstract Sociosexual behavior can be defined as behavior that entails the movements of sexual behavior (e.g., mounting) but occurs in wider social contexts (e.g., intercourse, play, reconciliation). An important goal of research on human evolution and health across the life course is to synthesize the limited data available on human infant and juvenile sociosexual behavior. To many, the subject of this chapter may seem like a non-starter, both because it is morally and religiously off limits to some, and also because of the conventional scholarly view that sexuality appears relatively suddenly at puberty. As we illustrate here, human sociosexual behavior has a start early in development, and an evolutionary perspective can highlight adaptive and health-related aspects of its patterning. Toward those ends, we first discuss the evolution of the human life history, highlighting the early ages of weaning and extended juvenile phases, and how those provide expanded opportunities for sociosexual play. We consider these aspects of the human life history in comparative context, drawing upon nonhuman primate examples of sociosexual development. We then review the scant available data on human subadult (including infants, nursing children, and weaned pre-adolescent juveniles) sociosexual behavior, also pointing toward health-related implications of this body of work.

4.1 The Evolution of Human Subadulthood, Learning and Play

> Beginning in the 1960s, [leading scholars] all made major claims about the importance of evolution for the study of human behavioral development. Lorenz and Tinbergen…saw in human development a logical practical application for their expertise on the apparent confrontation of nature and nurture. Skirting this ancient battlefield, they called special attention to the role of species-specific behavior and to the functions of play in the growth of competence.
>
> —Melvin Konner, *The Evolution of Childhood*

P.B. Gray (✉)
Department of Anthropology, University of Nevada, Las Vegas, NV, USA
e-mail: peter.gray@unlv.edu

M.H. McIntyre
Department of Anthropology, University of Central Florida,
4000 Central Florida Blvd, Orlando, FL 32816, USA

© Springer Science+Business Media New York 2017
G. Jasienska et al. (eds.), *The Arc of Life*, DOI 10.1007/978-1-4939-4038-7_4

Infancy is typically defined as the period from birth to weaning, and is followed by the *juvenile* stage (weaning to puberty). As Bogin (1999) has argued, humans wean their offspring at younger ages than expected based on the life histories of other primates, resulting in the emergence of an additional *childhood* stage of the life course. The term *subadult* can be used to embrace all of these prepubertal life history stages—infancy, childhood, and juvenility.

The evolution of human subadulthood has attracted considerable attention (Bogin 1999; Muehlenbein 2010). Compared with that of our ape relatives, a striking feature of the human life history is early weaning. While wild chimpanzees wean their offspring by around 4–5 years, and orangutans by 6–7 years, infants of hunter-gatherers are weaned by around 3 years, with weaning even earlier in other non-foraging societies (Hrdy 1999; Robson and Wood 2008). Another striking feature of the human subadult life history is the extended juvenile phase—the period of growth and development that precedes puberty and adolescence. This extension is accompanied by delayed onset of reproduction; presumably, the costs of this delay are outweighed by the fitness benefits of a longer developmental period. Recent data on the age at first reproduction for human hunter-gatherer females support the finding of a later onset of reproduction (e.g., around 18–20 years of age) for humans than for wild apes, such as chimpanzees (around ages 13) and orangutans (around age 15) (Robson and Wood 2008).

A variety of models have been advanced to account for these derived features of the human life history. Charnov (1993) sees an extension of all life history phases resulting from lower adult mortality, with the extended juvenile phase an incidental result of this overall extension. Bogin (1999) suggests that human alloparental care (care by individuals other than the mother) enhances lifetime female fertility by reducing interbirth intervals (time between successive births) and increasing subadult survival, incidentally yielding a childhood phase. Others focus on the adaptive benefits of an extended juvenile phase for learning essential ecological and social skills (see Geary 2010; Konner 2010), including skills needed to negotiate same-sex and mating relationships—all of which ultimately determine reproductive success.

Debates continue over the selective forces driving the evolution of earlier ages of weaning and extended juvenile phases in humans (including whether social learning is a cause or consequence of this evolution); these derived life history features undoubtedly have implications for the development of sociosexual behavior. Extended subadult phases provide more time to learn (Geary 2010; Konner 2010) locally relevant economic tasks (e.g., foraging skills) as well as social skills necessary for survival and reproduction. Through *play*, subadults can develop motor, cognitive, and emotional capacities during a developmentally sensitive phase that allows fine-tuning capacities needed for survival and reproductive success. Play entails opportunities for motivated practice and experience during a behaviorally plastic stage, without the high-stakes reproductive competition of adulthood.

Play that anticipates adult sociosexual behavior clearly should be important in an evolutionary context, and the development of sexuality, as distinct from other sex-typed behaviors, has obvious adaptive value. Young males of most mammal species are seen as needing to learn about male–male competition and to be sufficiently

Fig. 4.1 How children typically *don't* learn about sociosexual behavior. Baby reading Kinsey male volume (KI-DC: 44933)

motivated to perform basic mating behaviors when the opportunity arises. Young females are thought to be mainly learning about mothering and to be sexually receptive at the right times. As we learn more about the functions of adult sexuality, however, including infanticide avoidance and pair bonding, questions about the development of these behaviors become more interesting, particularly in the broader adaptive context for sex-typed and reproductive behavior. For more specific examples, we turn to comparative studies of nonhuman primate sociosexual development, and then consider human patterns of sociosexual development in greater depth (Fig. 4.1).

4.2   Nonhuman Primate Sociosexual Patterns

> Given that sociosexual patterns are found in all the great apes and are evident during infancy and juvenile life in orang-utans and gorillas, though much less commonly in adulthood, it is reasonable to suggest that sexual patterns might occur during play and other activities in human children.
>
> —Alan Dixson, *Primate Sexuality*, 2012

Nonhuman primates clearly engage in sociosexual behavior as subadults. As Alan Dixson's notes in his 2012 review of primate sexuality, "Sociosexual patterns of mounting, presentation, mutual embracing, and genital inspection or manipulation begin to develop during infancy in many monkeys and in the great apes." (Dixson 2012, p. 153). For example, male stumptail macaques play with each other's genitals; juvenile squirrel monkey males display their erect penes to others; and male rhesus monkeys engage in dorsal–ventral mounting of other males.

To understand sociosexual behavior of our recent ancestors, we can compare humans with the great apes—particularly chimpanzees and bonobos—who share a common ancestor around one million years ago. Although young chimpanzees begin sex play earlier and at higher rates than gorillas (Watts and Pusey 1993), bonobo juveniles engage in even higher rates of sexual contact (same- and opposite-sex) than do chimpanzees, paralleling the notorious differences between the two species in adult sociosexuality (Woods and Hare 2010). The shortage of relevant data on gorillas and orangutans makes it difficult to say more, other than that juvenile sociosexuality very likely occurred among our forebears—a pattern consistent with that in a wide range of primate species.

The nonhuman primate pattern of "sex-typed" behaviors generally parallels the sex differences commonly observed in humans, such as higher rates of rough-and-tumble play for males and offspring-play care for females (Cohen-Bendahan et al. 2005; Hines et al. 2004; Lippa 2005). Meredith (2012) recently summarized these patterns in wild nonhuman primates: "Juvenile females show more interest in infants than their male peers in many species…when sex differences in play are found, males play more frequently and more intensely than females (pp. 17–18)." In addition, a recent study of wild chimpanzees showed that when playing with a stick, juvenile females acted as if the stick were a baby (like a human child with a baby doll) more often than did juvenile males (Kahlenberg and Wrangham 2010). Among captive infant gorillas, males engaged in higher rates of social (e.g., wrestling) and object play than females, and both males and females preferred playing with male partners. These infant gorilla patterns parallel adult gorilla social behavior (Maestripieri and Ross 2004).

These specific patterns of primate subadult sexual and sex-typed behaviors are likely to have arisen through natural selection, and not only as inherited patterns constrained by our ancestry. Although few studies have been conducted on socially monogamous primates, Chau et al. (2008) noted that sex differences in subadult play tend to be found in polygynous species, such as squirrel monkeys, but are absent in more monogamous species such as common marmosets and cotton-top tamarins. This trend underscores the probable adaptive value of sex differences in juvenile sociosexual behavior: as a form of training for adulthood, where sex-specific behaviors are critical for reproductive success—for example, mothering in the case of females and fighting in the case of males.

4.3   Patterns of Human Sociosexual Development

[H]ugging and kissing are usual in the activity of the very young child, and…self manipulation of genitalia, the exhibition of genitalia, the exploration of genitalia of other children, and some manual and occasionally oral manipulation of the genitalia of other children occur in the two- to five-year olds more frequently than older persons ordinarily remember from their own histories. Much of this earliest sex play appears to be purely exploratory, animated by curiosity…

—Alfred Kinsey and colleagues, Sexual Behavior in the Human Male

Although data on human hunter-gatherers are sparse, there have been reports describing childhood sociosexual behavior among several different societies. A biography of Nisa, a !Kung woman from southern Africa, describes the ways children learn and practice sexual behavior (Shostak 1981). !Kung children sleeping near their parents hear them having sex at night. Girls engage in sex play with other girls, boys with boys, and boys and girls play together, often out of sight of adults. Typical sex play includes elements of coupling, affairs, and pretend intercourse, with play becoming more adult-like among older children. Among Aka foragers of Central African Republic, Barry and Bonnie Hewlett (Hewlett and Hewlett 2010, p. 114) note that, "Some [Aka] mentioned that sometimes children of the same sex (two boys or two girls) imitate parental sex while playing in camp and we have observed these playful interactions." And Frank Marlowe comments, regarding Hadza foragers in Tanzania that, "Hadza girls and boys begin "playing house" literally, building little huts, around the age of 7 or 8. There is some sex play when they enter the huts. Sometimes sex play among children occurs in full view of everyone; sometimes it is between two children of the same sex. Once, several Hadza and I watched two girls about 8 years old hugging and rolling around on the ground, clearly enjoying themselves in a sexual way. With increasing age, this sex play disappears; at least, it disappears from view." (Marlowe 2010, p. 168). Among Australian aborigines from Arnhem land, boys and girls pretend to engage in intercourse and play husband and wife, with these aspects of sex play becoming more adult-like the closer they are to adolescence (Berndt and Berndt 1951). Similarly rich anecdotes are available from horticultural societies.

Robert Suggs notes among 1960s Marquesan Islanders that, "Masturbation among males begins at about the age of three, or sometimes earlier.... At the age of approximately seven years, other forms of group sexual activity appear, which are heterosexual. Boys and girls, playing at "mother and father," will often place their genitalia in contact for brief periods. The girl either stands against a tree or lies supine on the ground, with the boy assuming the normal position for coitus. Contact is brief, accompanied occasionally by pelvic movement with much laughter. This activity is carried out in isolated areas where adults will not be apt to surprise the gathering." (Suggs 1966). In this account, age-related patterns of sociosexual play stand out, by which autosexual (e.g., masturbation) behaviors initially prevail, but are later complemented by more interpersonal sociosexual behaviors including play marriage. Generally, fewer data are available on girls' sociosexual behavior than boys', and there are important considerations of privacy between children and adults around engaging and observing sociosexual behaviors.

A classic cross-cultural survey by Ford and Beach (1951) focused on attitudes toward childhood sexuality using information coded from a culturally standardized data set. They identified three broad categories of societies: restrictive, semi-restrictive, and permissive. They documented considerable cross-cultural variation in how childhood sexuality was perceived, with some societies far more open to its expression and exploration than others. At one extreme, permissive societies included the Alorese of Indonesia, in which mothers would sometimes play with a nursing child's genitals, and young boys might masturbate in view of adults. At the other end of the spectrum, boys of the Murngin of Australia, a restrictive society were removed from their parents' dwelling to a bachelor or boys' house in order to prevent them from seeing their parents having sex. In addition to documenting and categorizing variation in childhood sexuality, Ford and Beach recognized that the way in which childhood sexuality was viewed within a society tended to be associated with restrictions on adult sexuality. That is, children's sociosexuality was structured similarly to that seen in adulthood.

The cross-cultural evidence shows that sociosexual behavior normally begins before puberty. This is a typical primate pattern: it takes time for individuals to learn all kinds of behavior, including sociosexual behavior suitable in a given social context for each sex. Sociosexual play provides practice and experience in behaviors that are later crucial for reproductive success. The pre-adolescent origins of human sociosexual behavior are also consistent with early physiological development of the sexual response: young boys are capable of erections, girls of vaginal lubrication, and both sexes of orgasm (without ejaculation) (LeVay and Baldwin 2009). One remarkable fetal ultrasound documented what appeared to be a boy masturbating, suggesting that some of the pleasure circuitry of the sexual response may develop even before birth (Meizner 1987). Moreover, as noted in the Hadza anecdote above and other research on children in the USA (Maccoby 1998), human childhood sex play has a context of long-term reproductive bonds (such as marriage). Sex play might be expected to develop in parallel with social skills associated with finding and maintaining a long-term partnership in humans; this is not the case for most other primates, especially our closest living relatives, chimpanzees and bonobos, consistent with the fact that they do not typically have long-term partnerships.

As a complement to cross-cultural accounts, in-depth quantitative studies of childhood sexuality have been undertaken in the USA and several European countries (Table 4.1). The first of these was launched by Alfred Kinsey, and based largely on interviews with adults asked to recall features of their childhood sexuality (Kinsey et al. 1948, 1953). While aspects of Kinsey's work remain controversial (e.g., non-random samples, ethical issues concerning the inclusion of data obtained from a man engaging in sex contacts with children), no better U.S. data on childhood sociosexual behavior have been obtained since. Among the important patterns found by Kinsey and colleagues was the finding that boys expressed greater motivation than girls to engage in sex play.

More recent large-scale US and European studies have attempted to characterize patterns of childhood sexuality by asking parents about their kids' sexual behavior. Since older children may be more aware of their behavior, and desire to shield it from parents' eyes, data from such studies more closely reflect parental observations than actual childhood sexual behavior (This methodological issue may help reconcile discrepant results in Kinsey's work with more recent Dutch and US studies based on parental reports: i.e., Mallants and Casteels 2008; Schoentjes et al. 1999). Nonetheless, results from other studies, some relying on individual self-report of children (rather than or parent, teacher, or doctor) support the following theme: "[T]here seems to be a gradual rise in masturbation in the prepubescent years, from around 10 % at the age of 7 to about 80 % at the age of 13." (Mallants and Casteels 2008:1113).

Table 4.1 Summary of in-depth quantitative studies of childhood sexuality in the USA and Europe

Sample	Key findings	References
U.S., mid-twentieth century: adults recalled their own childhood sexual behavior	~10 % of females reported sex play between ages 5–13; ~10 % of males reported sex play by age 5, increasing to ~35 % of males by age 11–13; ~2–5 % of females reported opposite-sex coital play across ages 5–13; 3 % of males by age five increasing to 12 % of males age 11–13 reported opposite sex coital play	Kinsey et al. (1948, 1953)
~1000 Dutch children aged 2–12: Based on parents' reports of children's sexual behavior	Frequency of sexual behavior decreased across ages 2–12, with similar rates between boys and girls Children exhibited increased sexual curiosity with age (e.g., the percentage of children asking about sexuality rose from 33 % aged 2–5 to 65 % aged 6–9 and 10–12)	Schoentjes et al. (1999)
~1100 US children: Based on parents' reports of children's sexual behavior	Percentage of children reported as touching their sex parts in public, at home, masturbating by hand or with a toy/object tended to decline across age groups (e.g., 16 % of girls aged 2–5 but 7 % of girls aged 10–12 were reported to masturbate by hand)	Mallants and Casteels (2008)
Swedish sample of 269: High-school senior students recalling childhood sexual experiences	Before age 13, over 80 % reported having engaged in autosexual (e.g., masturbation) and over 80 % recalled sexual experiences with another child	Larsson and Svedin (2002)

In addition to immediately influencing sociosexual behavior, a child's socioecology may impact adult sociosexual behavior. Draper and Harpending (1982) theorized that a father's presence early in a child's development channels adult sociosexual behavior toward less mating effort. Other models extend beyond a father's presence more generally to risks to reproductive success (as correlated with variables such as external mortality), and predict that children developing in higher-risk environments should undergo an earlier sexual debut and bias reproductive effort of reproductive effort toward actual mating effort (Chisholm 1999; Ellis 2004). There is some support for these models using data from US populations, but the international evidence largely refutes them (e.g., Schmitt 2005). Other individuals besides fathers with whom a young child spends considerable time (e.g., living with) may also be shunned as adult mates, as postulated by Westermarck (1923). There are also multiple lines of evidence supporting the idea that shared early-life experiences act against later sexual desire, consistent with an adaptive developmental process fostering incest avoidance (see Wolf and Durham 2004).

These age-related changes in sociosexual behavior suggest that children gain considerable practice by stimulating themselves and in sexual interactions prior to puberty, when the sexual stakes become much more significant. While boys and

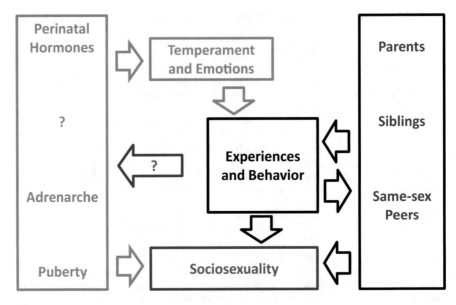

Fig. 4.2 A model of the development of human sociosexual behavior

girls appear to express a comparable sexual curiosity earlier in development (e.g., up to age 5), their sexual behavior differentiates as they grow older, with males engaging in higher rates of sex play than girls. These sex differences in sexual desire at young ages resemble those in adulthood, and may be attributable to mechanisms of sex differentiation (e.g., organizing effects of perinatal androgens on the brain: McIntyre and Hooven 2009). Girls and boys seem to engage in more adult-like sexual behavior the closer they are to puberty, as they prepare for the realities of sex and reproduction. The fact that older children seek to shield their sexual behavior from adults suggests that privacy is important for human sex (Gray and Anderson 2010). It may also be consistent with evolutionary theory about parent–offspring conflict, in the sense that it may be advantageous for kids to avoid parents' interference (Trivers 1974). Finally, sex play featuring couples (e.g., husband and wife) is consistent with adult human behavior, in which long-term bonds are the typical context for sexual behavior (e.g. Gray and Garcia 2013) (Fig. 4.2).

4.4 Development of Sociosexual Behavior

Androgens and estrogens play a central role in the pre- and perinatal development of sex differences in the reproductive tract. Later on, as puberty approaches, steroid sex hormones shape the development of sexually dimorphic traits, including differences in body size, shape, and growth patterns. These hormones are also implicated—although less firmly—in the development of the brain (even before birth) and of sex-differentiated behaviors in children (Table 4.2).

Table 4.2 Methods of studying biological sex differentiation of children

Method	Measurement of	Advantages	Disadvantages
Congenital adrenal hyperplasia	Prenatal androgen exposure in girls	Known, substantial hormone exposure	Genitalia may be masculinized (in addition to brain)
Amniotic fluid hormones	Prenatal steroid levels in both sexes	Most direct measure of normal variation	Amniocentesis for clinical purposes only
Pregnancy hormones	Prenatal steroid levels in both sexes?	Samples potentially more available and representative than amniocentesis	Probably little association between maternal and fetal hormones levels
Digit ratios (e.g., 2D:4D)	Mixture of prenatal androgens and estrogens?	Simple to collect in large community samples	Small and spurious effects may be common
Anogenital distance	Prenatal masculinization	Well validated in nonhuman studies	Less researcher interest in taking these measurements

A developing child's prenatally sex-differentiated brain is influenced by social contexts, including cultural expectations of gender differences in children's behavior, different learning models and availability of play partners. For example, the sex of older siblings can exert subtle effects on sex-typical behaviors. U.K. children who have older brothers have been shown to exhibit more male-typical behaviors, and children with older sisters more female-typical behaviors (Rust 2000). In other studies, boys who tend to shy away from male-typical social behaviors (e.g., rough-and-tumble play) are more likely to express a same-sex sexual orientation as adults. One possible explanation for this is that initial personality and emotional reactions shape an individual's engagement with the social world, which then shapes the expectations of his preferred friends and mates (see McIntyre and Hooven 2009).

One of the most notable social influences on sexuality is birth order (Poiani 2010). Having lots of older siblings, particularly older brothers, is the best predictor to date of a male having an adult same-sex sexual orientation, at least among the predominantly European and North American samples in which this subject has been researched. Similarly, Vasey and VanderLaan (2007) found that fa'afafine, Samoan men who have a strong feminine orientation and play alternative gender roles tend to have more older brothers (but they also tend to have more older sisters and more brothers overall).

There are various ways in which an association could arise between birth order and sexual orientation or gender role. A major proponent of the "fraternal birth order effect," Ray Blanchard, speculates that maternal immune responses underlie this pattern. Camperio-Ciani et al. (2004) have suggested that male homosexuality is linked pleiotropically to higher female fertility and larger families. One could as well imagine an effect of subtle differences in the social niches of boys with different numbers of older male siblings, with older siblings more able to aggress and coerce a younger one, and perhaps different pressures and expectations from parents.

As well as the children themselves, adults help create the social niche within which children's sex-typed behavior develops. Boys and girls both tend to be physically close to their mothers and other family members as infants. Those social links loosen as children grow older and more independent. As Maccoby (1998) notes from predominantly U.S. studies, girls around age 3 show preferences for same-sex relationships, as do boys about a year later. Sex segregation is more pronounced by mid-childhood. Cross-culturally, these mid-childhood transitions also tend to be marked by greater gender segregation in economic and social activities, as in adults. Girls tend to cluster with mothers, learning about household tasks and engaging in more childcare, while boys tend to be oriented more toward older males, including fathers, from whom they learn about status and economic activities. Several psycho-physiological mechanisms may also facilitate mid-childhood social behavioral changes, including adrenarche (Campbell 2006) and enhanced reasoning abilities (Konner 2010).

4.5   Synthesis and Conclusion

Patterns of human sociosexual development can be understood in the context of life history theory. A commonly held view is that human reproductive effort begins at puberty. However, the body of work reviewed here indicates that investment in mechanisms and experiences leading to successful reproductive effort begins early in development, proceeds through reproductive maturity, and is probably not strongly dependent on contemporaneous sex hormone levels. Moreover, human sex differences in sociosexual development suggest considerable allocation of time and energy in sex-specific juvenile patterns of "practice" mating and parenting effort. Girls (as in other primates) typically orient more toward offspring care, and boys (like other, but not all, primates, and excluding some socially monogamous ones) toward more rough-and-tumble play. These behaviors can be viewed as forms of sociosexual play that may have evolved as developmental preparation for behaviors with direct reproductive consequences later in the life history.

There are additional implications of the work reviewed here for sexual and public health. First, an appreciation of the complex interactions among factors influencing children's sociosexual development requires an understanding of its plasticity and variability. Much of that variability, in turn, is attuned to an individual's social context and environment, and presumably is geared toward phenotypically adaptive norms. Second, the shortage of opportunities for children to learn about sociosexual behavior through observational learning (in contrast to the case for other primates) may present novel cognitive and emotional challenges to children, particularly in societies that enforce more adult privacy. In American society, the boundaries between children and sexuality concern more than just privacy, and have strong moral and political dimensions.

From an ecological perspective, health depends both on the evolved functions of the body (physiological constraints) and current adaptive context for the individual. An evolutionary approach to understanding the sexual health of developing children can be employed to focus on two key questions: first, why have children evolved to be sexual at particular ages; and second, why might sexuality or asexuality of children be harmful in today's culture in a particular context at particular ages. These questions should be considered independently. In the case of behavioral health in general, and sexual health in particular, it can be difficult to strike a balance, since both the questions and answers have important social implications. People often want the socially or politically "correct" answer also to be reinforced as biologically "natural," even when there is little biological justification for this.

The implicit idea, especially in the USA, that all child sexuality is unhealthy (and the corollary that sexuality develops "instinctually" at puberty) also makes it more difficult for researchers or children themselves to discover what constitutes healthy or natural sexuality (Angelides 2004). If they lack clear information about the practical and emotional dimensions of sociosexual behavior, children are likely to be more vulnerable to developing poor psychosexual and behavioral health. Many parents would like to learn more from their doctors about normal child sexuality (Thomas et al. 2004). For both children and parents, the vacuum of ignorance is often filled with misinformation about sexuality from television and especially the Internet (Brown et al. 2002). While adults may realize that these are not accurate depictions of human relationships and experiences, children cannot make these judgments, and may have few alternatives.

In addition to the possible effects of privacy and misinformation on healthy knowledge about sexuality, there may be growing conflicts about sexual and behavioral health in societies with increasingly early puberty, on the one hand, and later ages of stable partnering on the other (Gluckman and Hanson 2006: see also chapters by Clancy and Reiches). In a modern cultural setting, sexual experience (and actual reproduction) that occurs too early may be seen as a threat to socially and economically adaptive life trajectories. These potential "mismatches" between biology and culture may contribute to adolescent body image dissatisfaction, depression, STIs, and pregnancies for which younger people are not cognitively prepared. As an illustration, in a nationally representative sample from Switzerland, early-maturing girls were more likely to be sexually active and to report body image concerns, while early-maturing boys were more likely to be sexually active, depressed, and to have attempted suicide (Michaud et al. 2006). When understood in evolutionary context, the sociosexual play of the earlier years of life must prepare young adults for the elevated sociosexual stakes of adulthood, and lifelong reproductive, social and economic success.

4.6   Study Questions

1. Suppose you were a chimpanzee attempting to decipher some of the species-, population-, and sex-specific patterns of human sociosexual development. What are some of the patterns you would find most noteworthy, and why?
2. Explain why the development of human sociosexual behavior represents neither "nature" nor "nurture," but a dynamic interaction of heritable influences and socioecological context.
3. Discuss three potential sexual or public health outcomes that stem from an understanding of the development of human sociosexual behavior.
   Helpful website: Kinsey Institute: http://www.kinseyinstitute.org/

References

Angelides S (2004) Feminism, child sexual abuse, and the erasure of child sexuality. Glq J Lesbian Gay Stud 10(2):141–177. doi:10.1215/10642684-10-2-141

Berndt RM, Berndt CH (1951) Sexual behavior in western Arnhem Land. University of California Press, Berkeley

Bogin B (1999) Patterns of human growth, 2nd Ed. Cambridge University Press, New York

Brown JD, Steele JR, Walsh-Childers K (eds) (2002) Sexual teens, sexual media: investigating media's on influence on adolescent sexuality. Lawrence Erlbaum Associates, Mahwah

Campbell BC (2006) Adrenarche and the evolution of human life history. Am J Hum Biol 18:569-589

Camperio-Ciani A, Corna F, Capiluppi C (2004) Evidence for maternally inherited factors favouring male homosexuality and promoting female fecundity. Proc R Soc B 271(1554):2217–2221

Charnov E (1993) Life history invariants. Oxford University Press, Oxford

Chau MJ, Stone AI, Mendoza SP, Bales KL (2008) Is play behavior sexually dimorphic in monogamous species? Ethology 114(10):989–998. doi:10.1111/j.1439-0310.2008.01543.x

Chisholm JS (1999) Death, hope and sex: steps to an evolutionary ecology of mind and morality. Cambridge University Press, Cambridge

Cohen-Bendahan CCC, van de Beek C, Berenbaum SA (2005) Prenatal sex hormone effects on child and adult sex-typed behavior: methods and findings. Neurosci Biobehav Rev 29:353–384

Dixson AF (2012) Primate sexuality: comparative studies of the prosimians, monkeys, apes, and human beings. 2nd Ed. Oxford University Press, Oxford

Draper P, Harpending H (1982) Father absence and reproductive strategy: an evolutionary perspective. J Anthropol Res 38(3):255–273

Ellis BJ (2004) Timing of pubertal maturation in girls: an integrated life history approach. Psychol Bull 130(6):920–958. doi:10.1037/0033-2909.130.6.920

Ford CS, Beach FA (1951) Patterns of sexual behavior. Harper & Row, New York

Geary DC (2010) Male, female: the evolution of human sex differences, 2nd edn. American Psychological Association, Washington, DC

Gluckman PD, Hanson MA (2006) Changing times: the evolution of puberty. Mol Cell Endocrinol 254:26–31. doi:10.1016/j.mce.2006.04.005

Gray PB, Anderson KG (2010) Fatherhood: evolution and human paternal behavior. Harvard University Press, Cambridge

Gray PB, Garcia JR (2013) Evolution and human sexual behavior. Harvard University Press, Cambridge

Hewlett BS, Hewlett BL (2010) Sex and searching for children among Aka foragers and Ngandu farmers of Central Africa. Afr Study Monogr 31(3):107–125

Hines M, Brook C, Conway GS (2004) Androgen and pyschosexual development: core gender identity, sexual orientation, and recalled childhood gender role behavior in women and men with congenital adrenal hyperplasia (CAH). J Sex Res 41(1):75–81

Hrdy SB (2009) Mothers and others. Harvard University Press, Cambridge

Kahlenberg SM, Wrangham RW (2010) Sex differences in chimpanzees' use of sticks as play objects resemble those of children. Curr Biol 20(24):R1067–R1068. doi:10.1016/j.cub.2010.11.024

Kinsey A, Pomeroy WR, Martin CE (1948) Sexual behavior in the human male. W.B. Saunders, Philadelphia

Kinsey A, Pomeroy WR, Martin CE (1953) Sexual behavior in the human female. Indiana University Press, Bloomington

Konner M (2010) Evolution of human childhood. Harvard University Press, Cambridge

Larsson I, Svedin CG (2002) Sexual experiences in childhood: young adults' recollections. Arch Sex Behav 31(3):263–273. doi:10.1023/a:1015252903931

Levay S, Baldwin JI (2009) Human sexuality, 3rd edn. Sinauer Associates, Inc., Sunderland

Lippa RA (2005) Gender, nature, and nurture, 2nd edn. Lawrence Erlbaum Associates, Mahwah

Maccoby EE (1998) The two sexes: growing up apart. Harvard University Press, Cambridge, Coming Together

Maestripieri D, Ross SR (2004) Sex differences in play among western lowland gorilla (*Gorilla gorilla gorilla*) infants: implications for adult behavior and social structure. Am J Phys Anthropol 123:52–61

Mallants C, Casteels K (2008) Practical approach to childhood masturbation: a review. Eur J Pediatr 167(10):1111–1117. doi:10.1007/s00431-008-0766-2

Marlowe F (2010) The Hadza: hunter-gatherers of Tanzania. University of California Press, Berkeley

McIntyre MH, Hooven CK (2009) Human sex differences in social relationships: organizational and activational effects of androgens. In: Gray PB, Ellison PT (eds) Endocrinology of social relationships. Harvard University Press, Cambridge, pp 225–245

Meizner I (1987) Sonographic observation of in-utero fetal masturbation. J Ultrasound Med 6(2):111

Meredith SL (2012) Identifying proximate and ultimate causation in the development of primate sex-typed social behavior. In: Clancy K, Hinde K, Rutherford J (eds) Building babies: primate development in proximate and ultimate perspective. Springer, New York

Michaud PA, Suris JC, Deppen A (2006) Gender-related psychological and behavioural correlates of pubertal timing in a national sample of Swiss adolescents. Mol cell endocrinol 254:172–178.

Muehlenbein M (2010) Human evolutionary biology. Cambridge University Press, New York

Poiani A (2010) Animal homosexuality: a biosocial perspective. Cambridge University Press, Cambridge

Robson SL, Wood B (2008) Hominin life history: Reconstructions and evolution. J Anat (212):455–458

Rust PCR (2000) Bisexuality: a contemporary paradox for women. J Soc Issues 56(2):205–221

Schmitt DP (2005) Sociosexuality from Argentina to Zimbabwe: a 48-nation study of sex, culture, and strategies of human mating. Behav Brain Sci 28:247–311

Schoentjes E, Deboutte D, Friedrich W (1999) Child sexual behavior inventory: a Dutch-speaking normative sample. Pediatrics 104(4):885–893. doi:10.1542/peds.104.4.885

Shostak M (1981) Nisa: the life and words of a !Kung woman. Vintage, New York

Suggs R (1966) Marquesan sexual behavior. Harcourt, Brace and World, New York

Thomas D, Flaherty E, Binns H (2004) Parent expectations and comfort with discussion of normal childhood sexuality and sexual abuse prevention during office visits. Ambul Pediatr 4(3):232–236. doi:10.1367/a03-117r1.1

Vasey PL, VanderLaan DP (2007) Birth order and male androphilia in Samoan fa'afafine. Proc R Soc B 274:1437–1442

Trivers RL (1974) Parent-offspring conflict. Amer Zool 14:249–264

Watts D, Pusey A (1993) Behavior of juvenile and adolescent great apes. In: Pereira ME, Fairbanks LA (eds) Juvenile primtaes: life history, development, and behavior. Oxford University Press, Oxford

Westermarck E (1923) The history of human marriage, 5th edn. Macmillan, London

Wolf AP, Durham WH (2004) Inbreeding, incest, and the incest taboo: the state of knowledge at the turn of the century. Stanford University Press, Palo Alto

Woods V, Hare B (2010) Bonobo but not chimpanzee infants use socio-sexual contact with peers. Primates 52(2):111–116

Chapter 5
Evolutionary Perspectives on Teen Motherhood: How Young Is Too Young?

Karen L. Kramer

Abstract Teen motherhood is the prevalent childbearing pattern in most traditional societies, as it likely was the case in the ancestral past. Yet teen pregnancy is associated with negative biological and social outcomes in the developed world. This contrast illustrates a question central to evolutionary medicine. How do adaptations to ancestral environments shape contemporary human health and behavior? Teen motherhood also exemplifies the important life history trade-off of whether to invest in current or future reproduction. Should a young teen continue to grow and mature or put her time and energy into starting a family? The age at which a young woman gives birth for the first time is ecologically sensitive to both her physical and social environment. Following an overview of the medical and social risks associated with teen pregnancy, this chapter presents a case study about the Pumé, a group of South American hunter-gatherers, to evaluate the costs and benefits of teen motherhood in a preindustrial environment. Results of demographic analyses show that the youngest of teen mothers ($\leq$14.3 years) have increased risk of infant mortality and lower lifetime fertility compared to older teens. However, young mothers gain no fitness advantage by delaying reproduction past their mid-teens. Cross-cultural comparisons suggest that childrearing practices rather than biological risks explain much of the discrepancy between traditional and developed societies in both the success of and attitudes toward teen motherhood.

5.1 Introduction

The topic of teen motherhood highlights a compelling question for evolutionary medicine: How young is too young to initiate reproduction? In developed societies, teen motherhood is considered a public health concern, commonly associated with poor pregnancy outcomes and negative social consequences, and has generated considerable public, media, and scholarly attention. Health risks to young mothers

K.L. Kramer (✉)
Department of Anthropology, University of Utah,
270 S 1400 E, Room 102, Salt Lake City, UT 84112, USA
e-mail: karen.kramer@anthro.utah.edu

© Springer Science+Business Media New York 2017
G. Jasienska et al. (eds.), *The Arc of Life*, DOI 10.1007/978-1-4939-4038-7_5

include a higher incidence of anemia, hypertension, obstetrical complications, and higher fetal, neonatal (Chen et al. 2007a, 2008; Gilbert et al. 2004; Jolly et al. 2000; Koniak-Griffin and Turner-Pluta 2001; Mahavarkar et al. 2008) and maternal mortality (Conde-Agudelo et al. 2005). Young mothers also are more likely to be unmarried, socioeconomically disadvantaged, and at greater risk of cigarette, drug, and alcohol use (Chen et al. 2007b; Debiec et al. 2010; King 2003). These trends paint a grim picture for teen mothers.

From an evolutionary perspective, if young motherhood poses significant morbidity and mortality risks for mothers and infants, natural selection is expected to favor later ages at first birth. Yet, in traditional societies, young women routinely give birth in their teens, as humans most likely did throughout the past. In traditional societies, young motherhood is sanctioned, encouraged, and does not have negative social repercussions. How can these two, very different societal views of teen motherhood be reconciled?

This chapter situates teen motherhood in comparative perspective by first presenting an overview of the clinical literature, followed by a summary of the main biological and social factors that affect the onset of female reproductive maturity and variation in age at first birth. Life history theory is then introduced as a useful framework for thinking about the costs and benefits associated with teen motherhood in different socio-ecological contexts. I then present a case study of early childbearing based on demographic analyses of the Pumé, a group of South American hunter-gatherers, to address the fundamental biological questions: How young is too young to initiate reproduction? And are risks substantially reduced if mothers delay childbearing until their 20s? The Pumé findings are then discussed in light of their relevance for understanding the contrasting views of teen motherhood in developed versus traditional societies.

5.2 Clinical Overview

Most of what is known about the health consequences of teenage pregnancy comes from clinical studies conducted in modern industrialized societies where multiple social factors such as access to health care and marital, socioeconomic, and educational status can affect pregnancy outcomes in complex ways. For example, the biological relationship between teen pregnancy and low infant birth weight—a primary predictor of infant morbidity and mortality—is difficult to isolate because various lifestyle risk factors (cigarette, drug and alcohol use, inadequate diet, uterine infection, and low prepregnancy weight) often present adverse effects irrespective of maternal age (Akinbami et al. 2000; Kramer et al. 2000). In studies that do control for social factors, the effect of maternal age has generated mixed results. While some studies have found little residual effect of maternal age on labor and delivery outcomes (Geronimus 1987; Lee et al. 1988; Makinson 1985), others have come to the opposite conclusion, reporting a higher risk of negative pregnancy outcomes for teenage mothers independent of known social confounds

(Chen et al. 2007b; Fraser et al. 1995; Jolly et al. 2000; Olausson et al. 1999; Phipps and Sowers 2002).

Two main reasons likely account for this ambiguity. First, although younger and older teens are in different stages of development, clinical studies generally group teens aged 15–19 years into one homogenous age category. Studies that have distinguished age-specific effects have found that adverse pregnancy outcomes are far more pronounced in very young teens (Forrest 1993; Fraser et al. 1995; Lancaster and Hamburg 2008 [1986]; Olausson et al. 1999; Satin et al. 1994), whereas risks tend to level off among older teens and resemble those of adults (Phipps et al. 2002; Phipps and Sowers 2002). In other words, grouping teen mothers into one age category likely underrepresents the negative consequences to younger teens and overstates risks to older teens. Second, age-specific health risks for teen mothers are more sensitive to gynecological age (time since onset of menarche) and skeletal age than to chronological age per se. For example, most teens continue to grow and do not have fully developed reproductive organs or hormonal cycles for 2–3 years following menarche (Stevens-Simon and McAnarney 1995). Reflecting this, conception within 2 years of menarche is associated with elevated risks of preterm birth (Stevens-Simon et al. 2002). Yet, despite risks being closely associated with gynecological age and reproductive immaturity, most clinical studies report results in terms of chronological age.

5.3 Female Reproductive Maturation

The biological capacity for a young woman to conceive and deliver a child is mediated by three developmental processes: the pace of juvenile growth, age at menarche, and the duration of subfecundity following menarche. Adolescent girls typically accomplish most of their statural growth before menarche can occur (Ellison 1981; Ellison 2001). During the subfecund period that follows, bi-iliac breadth reaches adult dimensions, and surplus energy (above maintenance requirements) is reallocated from skeletal growth to fat storage as ovarian function matures. Although this sequence of events is highly canalized, the pace and timing can vary widely across individuals and populations.

Although genetic, social and intergenerational effects account for some of the variation in female maturational pace, (Ellison 1990; Ellison 2001; Gillett Netting et al. 2004; Jasienska and Ellison 2004; Ulijaszek 1995; Worthman 1993), energy availability constitutes a prominent underlying factor. For example, although age at menarche is highly heritable (Towne et al. 2005; Perry et al. 2014), average age at menarche has decreased in many developed nations over the last century (Ellis 2004; Garn 1987; Nichols et al. 2006). With some variation across cohorts, among UK-born women, menarcheal age declined from a mean age of 13.5 years in the early twentieth century to 12.3 years in the late twentieth century (Morris et al. 2011). This decline in menarcheal age, known as the secular trend, has been attributed primarily to improved nutritional conditions (Parent et al. 2003). While average age at menarche appears to have stabilized in developed nations at around 12.5 years, of growing concern is the rising number of girls who reach menarche at a precocious age (before the age of 9, although specific age may vary depending on the study), a trend linked to increased childhood obesity and downstream pathologies (Currie et al. 2012).

Within populations, slower and later reproductive maturity is associated with poor childhood conditions (Ellis 2004; Eveleth and Tanner 1990; Foster et al. 1986; Garn 1987; Riley et al. 1993), while faster and earlier maturity has been related to improved childhood health and nutritional status (Ellis 2004; Eveleth and Tanner 1990; Foster et al. 1986; Garn 1987; Riley et al. 1993). Various social stressors, such as emotionally difficult childhoods and family dysfunction, also can have accelerating effects on developmental pace (Belsky et al. 1991; Boyce and Ellis 2005; Ellis et al. 2003; Kim et al. 1997), whereas stable high-quality social environments have been linked to later maturity (Chisholm et al. 2005; Coall and Chisholm 2003; Hulanicka et al. 2001). For example, urban Australian women who experienced troubled family relations early in life (under the age of 10) reached menarche at a significantly younger age (varying between 4 and 13 months depending on stress measure) than women who had less stressful young lives (Chisholm et al. 2005). Father absence, in particular, has been associated with early menarche and teenage pregnancy (Bogaert 2005; Ellis et al. 2003; Quinlan 2003). Although the psychophysiological mechanisms that might advance or delay reproductive maturity remain unclear, the connections being made between prenatal conditions, infant birth weight, and the timing of reproductive development may reveal new insight into the role of intergenerational and epigenetic effects (Adair 2001; Koziel and Jankowska 2002; Kuzawa 2005; Kuzawa and Sweet 2009; Meade et al. 2008).

5.4 Variation in Age at First Birth

In developed societies, the period between the onset of reproductive maturity and age at first birth may be considerable, often regulated through birth control rather than restricted sexual activity. Average age at first birth in developed societies ranges from 25.1 years in the USA to 29.9 years in Canada (United Nations Economic Commission for Europe 2007). Teen motherhood is viewed as a public health concern and generally discouraged throughout the developed world. In contrast, in natural fertility societies (in which parous-specific fertility control does not occur), teen motherhood is the pervasive childbearing pattern, often encouraged and endorsed as a societal norm. The average age at first birth for a large sample of traditional societies, including forager, horticultural, and agricultural populations, ranges from 15.5 years among the Pumé to 25.8 years among the Gainj (Fig. 5.1).

In natural fertility societies, a young woman's exposure to conception is usually mediated through marriage rules. In some societies girls may marry prior to or shortly following menarche and give birth to their first child soon after reaching

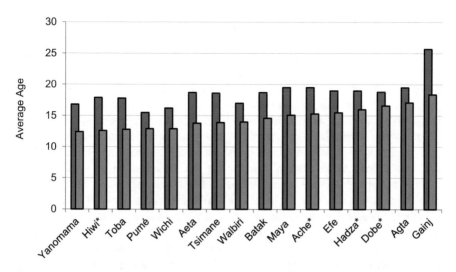

Fig. 5.1 Cross-cultural average age at menarche (*gray bars*) and first birth (*blue bars*), sorted on age at menarche. *Societies on which the human norm of 19.5 is typically based. Sources: Data reported for all available traditional small-scale societies: *Yanomama* (Early and Peters 1990); *Toba, Wichi* (Valeggia and Ellison 2004); *Pumé* (Kramer 2008); Aeta (Migliano et al. 2007); *Hiwi, Tsimane*, Walbiri, Efe, Hadza (Walker et al. 2006); Batak (Eder 1987); Maya (Kramer 2005); *Ache* (Hill and Hurtado 1996); Dobe Kung (Howell 2000 [1979]); Agta (Early and Headland 1998); Gainj (Wood 1992)

reproductive maturity. In other societies, girls may be sequestered, their fidelity closely guarded following menarche, with marriage and childbearing delayed for a period. In most cases, married women will live among close relatives throughout their lives and over the course of their reproductive career will bear 6–8 children, many of whom may not survive. Although women in traditional societies are not typically exposed to the lifestyle risks that can compromise teen pregnancy outcomes in developed societies, they tend to work physically hard for a living, may be food limited, and often have no access to prenatal health care.

5.5   Life History Framework

Life history theory is a useful framework to consider the optimal age to initiate childbearing under different circumstances. Women who start having children too young may compromise their own growth and potentially jeopardize their health as well as fetal development and offspring survival. Women who delay childbearing too long have shorter reproductive careers and potentially reduce the total number of offspring produced in a lifetime. As such, the timing of first birth represents a trade-off between the costs and benefits associated with current versus future reproduction (Clutton-Brock 1988; Stearns 1989). Age at first birth is influenced by extrinsic mortality (Charnov 1992; Ernande et al. 2004; Gadgil and Bossert 1970). In high-mortality environments, females may initiate reproduction early rather than risk not surviving to a later age. In contrast, a female is more likely to delay reproduction in low-mortality environments, especially if growing for longer periods and having a larger body size is beneficial. Empirical studies across a number of taxa confirm that age at first birth is sensitive to background mortality rates. For example, studies have shown that Northern cod start to mature at earlier ages and smaller body sizes in response to increased subadult mortality resulting from fishing practices (Olsen et al. 2004). Among adult Tasmanian devils, the development of a widespread fatal cancer has led to a dramatic increase in the proportion of early-maturing individuals (Jones et al. 2008). In humans, several studies indicate that girls growing up in high-mortality environments with low life expectancy tend to mature more quickly than girls growing up in safe, low-mortality environments (Chisholm 1999; Cooper et al. 1996; Ellis et al. 2009; Hill and Kaplan 1999; Ibáñez et al. 2000; Kramer et al. 2009; Lancaster et al. 2000; Walker et al. 2006). For example, the Hiwi and Pumé, two groups of foragers who live in a high-subadult mortality environment in western Venezuela, grow quickly during the juvenile period and reach menarche at a relatively young age: 12.6 years on average for the Hiwi (Walker et al. 2006) and 12.9 years for the Pumé (Kramer et al. 2009). Thinking about age at first birth as a life history trade-off between current and future reproduction offers insight into how teen motherhood varies in relation to different socio-ecological conditions. The Pumé foragers of South America are an ideal population to use a life history framework to investigate the biological costs and benefits associated with teen motherhood. Women in Pumé society have no access to education or health care and are not stratified by variation in food availability or other social factors known to affect birth outcomes in developed societies. Furthermore, teenage marriage and pregnancy are not stigmatized or associated with negative social consequences. These factors allow the fitness consequences of teen motherhood to be more clearly observed and evaluated. As in many hunter-gatherer societies, Pumé women marry young and usually initiate conjugal relations soon after reaching menarche, with 95 % of girls married by age 15 (Kramer 2008). Given their menarcheal and marriage pattern, the Pumé not surprisingly have a relatively high teen birth rate (defined as the number of births to adolescent girls ages 15–19 per 1000 adolescent girls ages 15–19) of 195 births per 1000 teens ages 15–19. This compares to 110 births per 1000 teens for Bangladesh, which has one of the highest national teen birth rates, 90 for Venezuela (Population Reference Bureau 2007), and 23.6 for the USA in 2013 (U.S. Department of Health and Human Services 2015).

From a life history perspective, evaluating the adaptive value of teen motherhood for the Pumé rests on addressing two central questions: (1) Does the risk of infant mortality vary with maternal age at first birth? (2) What are the lifetime fitness consequences of teen motherhood? In particular, if pregnancy at a young age poses significant consequences for lifetime reproductive success, then older first-time Pumé mothers should have lower infant mortality and/or higher lifetime fertility than younger first-time Pumé mothers.

5.6 Pumé Case Study

5.6.1 Study Population

The Savanna Pumé live in a remote area of south-central Venezuela on a low-lying plain (llanos) drained by tributaries of the Orinoco River. They have no access to modern amenities such as well water, electricity, health clinics or schools. None of the study communities can be reached by permanent road or has access to market goods. Nonlocal goods (cloth, metal pots, machetes) are obtained through trade with the River Pumé, a related group who live along the major rivers and transportation routes into the region. The Savanna Pumé exchange arrow cane and other raw materials for these goods, which are well worn by the time they reach the interior. Savanna Pumé women do not have access to market foods, health care, or prenatal intervention. Although the Pumé are food limited, adult height falls within the normal range for other native South American populations (Holmes 1995; Salzano and Callegari-Jacques 1988).

The Savanna Pumé move their camps 5–6 times throughout the year in response to seasonal changes in rainfall patterns. During the 6-month dry season, related nuclear families live in ephemeral brush-shade camps adjacent to streams and lagoons to be close to water and fish. During the wet season when the *llanos* flood, camps are moved to higher ground and families aggregate in more substantial thatch houses. Fish become dispersed and are difficult to locate, and the subsistence base shifts to small game, wild roots, and a small amount of cultivated bitter manioc (Greaves and Kramer 2013). The nutritional ecology of the Savanna Pumé is described at length elsewhere (Kramer and Greaves 2007). Nutritional and epidemiological stress, extreme in some years, is most pronounced during the wet season when relatively low food returns are exacerbated by increased exposure to mosquitoes and infectious disease (Barreto and Rivas 2007). The most prevalent epidemiological risks include malaria, respiratory infections, tuberculosis, giardia, amoebiasis, and parasitic infections. Although some older individuals have been immunized, primarily for small pox, very few Savanna Pumé children have been vaccinated in the last 15 years.

Serial monogamy is the predominant marriage pattern for the Pumé, although 20 % of adults have been polygynously married at some point during their lives. If an extramarital affair occurs, it usually results in divorce and remarriage. Women generally enjoy friendly and supportive marriages and have autonomy in decision-making, including when and whom they marry. Although marriage is often arranged by parents, young women are not obliged to accept these matches. Young couples often live with the wife's family for some years, and either partner can instigate divorce. Although Pumé girls may marry and reside with a spouse before menarche, coital relations are not initiated until later, usually shortly after menarche when girls are ready. No evidence exists of coercive marriage or sexual activity, an important consideration given the early age of marriage.

Savanna Pumé girls grow up in a high-mortality environment, marked by distinct seasonal fluctuations in food availability and harsh epidemiological conditions, with no access to health care, immunization, or supplemental food programs. Both childhood and adult morality are high (see fertility and mortality section below) and life expectancy at birth is low ($e_0 = 27$). This affects the age structure of the population. The average age of the population is 24.01 years, and 39 % of the population is under the age of 15. While this indicates that the Pumé are a relatively young population, the age structure also reflects the high child mortality (Kramer and Greaves 2007).

5.6.2   Data Collection

The information reported here builds on a project begun in the 1990s by archaeologist Russell Greaves who was researching the daily lives of hunter-gatherers. A decade later, we pooled our efforts and initiated a follow-up project focused on the demography and life history of the Pumé. One of the first steps in this project was to establish the ages and relatedness of individuals. This information is necessary to build accurate censuses (age, sex, names, and family relations), reproductive histories, and fertility and mortality profiles. Because first menses and first birth are discreet, life-changing events important to the lives of young girls and the community, these events are readily recalled by informants. However, most small-scale societies do not mark the passage of time in calendar dates or keep vital (birth and death) records. Additionally, among the Pumé as many hunter-gatherers, people address each other by kin terms, not by name. To overcome these challenges, anthropologists have developed a number of methods (see Box 1) for constructing reliable ages and dates (Hill and Hurtado 1996; Howell 2000 [1979]; Kramer 2005, 2011).

The data presented here were collected from three Savanna Pumé hunter-gatherer communities ($n = 235$) from 2005–2007. Many of the young women who were coming of age during the 2005–2007 field seasons had been born during the first phase of Pumé research in the 1990s, and their birthdates were known. Thus while the sample of young women is small, birth, first menses and parturition dates are known to the month and year and in some cases to the day (Kramer 2008; Kramer and Greaves 2007). Furthermore, although Venezuela does not regularly collect censuses in this region and rarely do these include more than a head count, we were fortunate in that anthropologists and demographer Roberto Lizarralde had spearheaded a Native census project and collected censuses among the Pumé several times throughout the 1980s and 1990s. These records were invaluable for anchoring the ages of most individuals who were in their adolescence and older during the 2005–2007 field study.

5.6.3 Main Variables

5.6.3.1 Fertility and Mortality

Two methods were used to estimate fertility for Pumé women. The *period* total fertility rate (TFR) is a cross-sectional measure, calculated as the sum of age-specific fertility rates at a given point in time. The *cohort* total fertility rate is a longitudinal measure, calculated as the average number of live births for a cohort of women who have completed their reproductive career, and relies on retrospective reproductive histories. The period and cohort TFR for Pumé women are 7.41 and 7.40, respectively (Kramer and Greaves 2007). The similarity of these estimates suggests that the age pattern of childbearing has changed little over recent decades (Preston et al. 2001). In addition, the relatively high TFR for Pumé women suggests that energy balance is sufficient to support conception and pregnancy despite seasonal undernutrition.

High infant and child mortality rates also characterize the Pumé: 35 % of live births do not survive infancy (the infant mortality rate = 346, the number of infant deaths per 1000 live births), and almost 45 % do not survive to reproductive age (Kramer and Greaves 2007). All Pumé mothers experience at least one infant death by their fifth parity, with 40 % of mothers ages 15–25 having had a child die in infancy by their second parity. (Infanticide has never been noted in the Pumé ethnographic or demographic literature.) Of the few hunter-gather groups for whom infant mortality has been reported, the Pumé rate is similar to those of the Agta, Asmat, and Mbuti and higher than those of the Hadza, !Kung, and Ache (Hewlett 1991; Pennington 2001). Savanna Pumé infants and young children are particularly challenged to live through their first wet seasons. Women can lose up to 8 % of their body weight during this time of year. While mother's milk quality is buffered in many nutritionally stressful circumstances, under extreme restriction, it can lead to maternal depletion and compromise lactating infants (Sellen 2000, 2006). Among the Pumé, wet season maternal depletion coupled with high pathogen loads likely exacerbate infant stress and contribute to the high levels of morbidity and mortality.

5.6.3.2 Age at First Birth

During reproductive history interviews, Pumé women were asked about their current reproductive status—whether they had experienced their first menses or had given birth. These cross-sectional data indicated what proportion of females in the current population had reached menarche or had given birth to their first child at each age. Based on this cross-sectional sample, the probability that females ages 10–25 (*n* = 27) had ever given birth can be modeled, giving a median age at first birth of 15.5 years (estimated at the .5 probability; Kramer 2008). To take advantage of the available data on women's reproductive events over the last 20 years, a mixed sample of longitudinal and retrospective data can also be used to model the probability that a woman has given birth to her first child. Based on this sample ($n = 38$ females ages 10–22), average age at first birth occurs at 15.3 ± .937 (SD) years. Calculating age at first birth using both methods assures that the strength of the results rests on more than one estimate.

Although Pumé mothers begin their reproductive career relatively early compared to other foragers, their average age at first birth is biologically reasonable given a subfecund period of about 2 years and a median age at menarche of 12.87 ± 1.02 (SD) based on the longitudinal sample ($n = 16$; Kramer 2008). Age at menarche for Pumé girls also falls within the range of normal variation for other traditional native South Americans (Walker et al. 2006).

5.6.4 Consequences of Age at First Birth on Infant Mortality and Lifetime Fertility

To determine the extent to which age at first birth is associated with the risk of infant mortality, first-time Pumé mothers ($n = 44$) were grouped into three age categories. Mothers whose age at first birth fell below one standard deviation of the mean (15.29 ± .937) were designated as early reproducers (11.8–14.34 years; $n = 12$); first-time mothers within one standard deviation of the mean were designated as average reproducers (14.35–16.23 years; $n = 16$) and those above one standard deviation of the mean as late reproducers (16.24–22.96 years; $n = 16$). Maximum likelihood estimates for the odds of a child dying showed no significant difference in infant mortality between average and late reproducers (odds ratio = .523, $p = .2760$). However, early reproducers were 3.7 times more likely to lose their firstborn compared to average and late reproducers ($p = .0134$). These results indicate that the youngest of Pumé mothers (under 14.34 years old) had a significantly elevated risk of infant mortality in their first parity.

One benefit for mothers initiating childbearing at a young age is that they will have a longer reproductive career and potentially greater completed lifetime fertility compared to mothers who delay and start their reproductive career at older ages. Since Pumé mothers on average bear 6–8 children over the course of their reproductive careers, does the elevated risk of infant mortality in the first parity for the youngest of Pumé mothers have an impact on total lifetime fertility? To address this question, a model of surviving fertility—the net outcome of fertility and parity-specific mortality probabilities—was calculated for a typical early, average, and late reproducer over the course of her reproductive career. As shown in Fig. 5.2, mothers who initiated childbearing in their early teens had lower surviving fertility at each parity compared to mothers who initiated childbearing in their mid- to late teens. To achieve the same level of surviving fertility as older first-time mothers, and thus recoup the high risk of infant mortality associated with the first parity, the youngest first-time mothers must bear an additional child—an unlikely outcome since few Pumé women bear eight children.

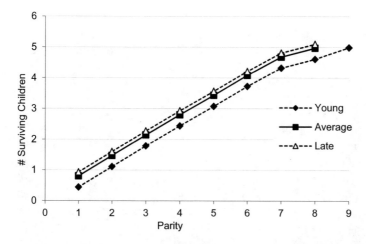

Fig. 5.2 Lifetime surviving fertility given the level of mortality at each parity for Pumé females ($n = 44$) who are young, average, and late age-at-first-birth reproducers. Surviving children calculated from empirical probabilities of infant mortality at each parity for young, average, and late reproducers

In sum, the mortality and fertility outcomes described above indicate that initiating childbearing in their mid-teens is the optimal reproductive strategy for Pumé women, enabling them to minimize the risk of infant mortality while maximizing their potential lifetime fertility. First-time mothers who are in their early teens pay a significant fitness cost both in terms of higher infant mortality and lower lifetime fertility. Mothers who delay first reproduction until their late teens, however, appear to gain no significant fitness benefit in terms of lifetime reproductive success.

What effect does early childbearing have on mothers themselves? Maternal depletion and the high energetic cost of lactation likely compromise a mother's immune and somatic expenditures, especially during the wet season when many women are nutritionally stressed. The cost to younger versus older women, however, is unknown. A meta-analysis of maternal mortality from 188 countries found that maternal mortality (number of maternal deaths per 100,000 live births) ranges from a high of 956.8 (or .956 %) in South Sudan to 2.4 (or .0024 %) in Iceland (Kassebaum et al. 2014). Given Pumé conditions, a high maternal mortality rate is likely, but cannot be calculated because of the small population and the relative infrequence of maternal mortality. Age-related risks across populations suggest that maternal mortality is greatest both for the youngest of mothers [ages 15 and younger (Conde-Agudelo et al. 2005; Harrison 1985)] and the oldest of mothers (Kassebaum et al. 2014). This age pattern is consistent with the elevated risk of infant mortality among the youngest Pumé mothers.

Box 1: Aging a Population Without Birth Records
Small-scale traditional societies such as the Pumé often use kin terms, not names, to address each other. Learning these terms is the first step that an anthropologist takes in order to hold basic conversations, form relationships and develop trust with informants. Among the Pumé very specific terms are used to reference older and younger individuals. These kin terms are used to establish birth orders and crosscheck relative ages within sibling groups. During a household interview we gathered all members together, including parents, children and older adults, which creates a forum for multiple relatives to talk about kin relations and, hopefully, agree on the sequence of births and deaths. As anyone growing up in a large family has experienced, children often know their ages better than their parents do. People also tend to list their children and siblings from eldest to youngest, which can be used to corroborate birth orders within sibships from multiple individuals. In some societies cohort memberships are recognized through various initiation rites and social ceremonies. In societies without calendar dates, time is often marked by the regular occurrence of natural or social events. For example, young Pumé children's ages can be recalled reliably in moon counts, while older children's ages can be recalled in dry and wet season counts. Sometimes a critical event such as a memorable flood, drought, epidemic or war can be used to reference dates. While it is usually not feasible to construct ages and dates with precision to the day, the methods described above often allow an anthropologist to obtain accurate ages and dates to the year and month.

5.7   Discussion

5.7.1   Childbearing Versus Childrearing

Teen first birth appears adaptive for the Pumé. Successful motherhood is contingent not only on supporting pregnancy and lactation during infancy, but also providing food and childcare during the long human juvenile period. In the case of the Pumé, young first-time mothers are less efficient food producers and less skilled caretakers than older first-time mothers (Kramer et al. 2009). For example, when foraging for tubers, an important food collected exclusively by women and comprising 40 % of the Pumé diet, young Pumé mothers ages 15–24 have significantly lower food return rates (amount of food collected per time spent collecting the food item) compared to older Pumé women (Kramer et al. 2009), a pattern also documented for other foragers (Hawkes et al. 1997; Hurtado et al. 1992).

Early childbearing is possible in traditional societies for certain reasons: young motherhood rarely occurs outside the context of marriage, teen mothers also live in

Fig. 5.3 Three generations of Pumé females. Photo credit: RD Greaves

extended and multigenerational families and are tied into extended kinship networks. Much of the care for very young children comes from the extended family (Fig. 5.3). Across traditional societies, infants receive about 50 % of their childcare from someone other than their mother (Kramer 2010), with the balance coming from female siblings (13–31 %), grandmothers (1–14 %), and fathers (<1–16 %). Young mothers in traditional societies tend to live, work, and learn adult skills in their familial or extended household and spend their time in domestic work or food production (gardening or foraging) activities that can be easily combined with childcare. Under these circumstances, conflicts so common in developed societies between spending time at work and taking care of children are more easily resolved (Fig. 5.4).

In addition to social support and childcare, teen mothers in traditional societies receive help from others through resource provisioning. The metabolic risks of survival and the high energetic costs of reproduction are often mediated through labor cooperation and food sharing, a distinction from most other animals referred to as pooled energy budgets (Kramer and Ellison 2010; Kramer et al. 2009). Pooling labor and sharing resources mean that some individuals, often pregnant and lactating women and young children, may produce fewer calories than they consume during certain life stages, while others, namely older children and adults, may produce more calories than they consume. For most mammals and primates (except for cooperative breeders), the energy required for postweaning growth, survival, and reproduction depends entirely on the individual's own ability to harvest calories

Fig. 5.4 Pumé woman combining work and childcare. Photo credit: RD Greaves

from the environment. Females only initiate reproduction after they have become self-sufficient foragers and remain on their own to acquire the extra calories needed to support pregnancy and lactation. The evolution of human cooperation and food sharing, however, permits the economics of childbearing to be independent of the biology of childbearing. While the timing of first birth is biologically constrained by reproductive maturation, because women receive help from others, it is flexible in terms of economic maturation and the ability to pay the costs of childrearing, an interesting distinction of humans from other mammals.

Pooled energy budgets also have implications for unraveling some of the complexities of life history trade-offs involved in the timing of first birth. The extent to which others can reduce a growing girl's activity load and/or preferentially feed her (Gillett Netting et al. 2004; Kramer et al. 2009) influences the amount of surplus energy available to complete statural growth, reach menarche, and shift energy allocation to reproduction. Because pubescent Pumé girls receive food from others, they can minimize energy expended on foraging tasks and other related activities. This, in turn, maximizes the energy available for growth and building fat stores in preparation for reproduction (Kramer and Greaves 2011). Pooled energy budgets allow young women in traditional societies to reproduce when they can biologically support childbearing, yet well before they can socially and economically support childrearing.

5.7.2 Developed Societies

A Pumé teenager and a twenty-first-century American teenager become mothers under very different circumstances, which affect their success as young parents. Teen motherhood in developed societies occurs in the context of an economy in which the ability to provide for children is generally contingent on the market value of wage-earning skills often learned formally in school. In these societies, teen motherhood typically incurs the opportunity cost—the value of a foregone activity—of not completing formal education or training. Delaying or foregoing education or training can have lifelong consequences on a teen mother's future ability to provide for herself and her children. Furthermore, demographic trends associated with modernization have greatly reduced kin-based childrearing networks. As average completed family size shrinks, for example, the number of siblings, cousins, aunts, and uncles available to help raise children diminish not only for current generations but also in subsequent generations (Kramer and Lancaster 2010). As the time between generations lengthens with later ages of marriage and childbearing, and families become more prone to geographic dispersion in pursuit of economic opportunities, kin networks diminish further and become more fractured. Compared to traditional societies, these demographic consequences of living in a developed nation can leave teen mothers more vulnerable and isolated, without kin networks and childrearing support. The US teen birth rate has substantially declined by 51 % over the past 20 years, but remains one of the highest rates in the developed world at 23.6 births per 1000 teens ages 15–19 (U.S. Department of Health and Human Services 2015; values may vary depending on reporting institution; e.g., see the Guttmacher Institute 2014). Most teen births (73 %) are to older teens ages 18–19, and the majority (89 %) occur outside marriage. Based on data from 2012, an estimated 1 in 8 teens (12.5 %) will give birth before their twentieth birthday. Teen birth rates vary by geographic region and ethnicity, with the highest rates reported among Hispanics (46.3 births per 1000 teens ages 15–19) and Blacks (43.9) and in the southern states, with the lowest rates among Whites (20.5) in the northeast.

Besides these demographic differences, studies of US Blacks have shown that social networks available to teens, especially from matrilineal kin, can significantly affect pregnancy outcomes and the success of young motherhood (Geronimus 2003; Geronimus et al. 1999; Lancaster and Hamburg 2008 [1986]). The relative risk of neonatal mortality, for example, is lower for Black teen mothers ages 17–19 compared to older mothers ages 24–29, and substantially lower compared to White mothers in the same age ranges (Geronimus 1986, Table 2). This difference has been attributed to Black teens receiving greater social and economic support from their matrilineal kin compared to older mothers and White teens when paired age for age. Black teens received more familial support because their own mothers were more likely to be alive and healthy compared to women who waited until their 20s to initiate childbearing (Geronimus 1996a). In addition, because of increasing health deterioration among older mothers, Black teens also showed lower incidences of low birth weight babies and other complications (Geronimus 1996b).

5.7.3   *Implications*

Evolution has produced a universal human life history pattern whereby females typically reach reproductive maturity before they are socially and economically competent caretakers. With this in mind, the social environment—rather than biological age per se—helps explain the cross-cultural differences between traditional and developed societies in the prevalence of and attitudes toward teen motherhood. Except for the youngest of teens childrearing practices, especially the presence or absence of kin-based support networks, largely determine the success of teen motherhood. In developed societies, there is rising concern about the widening discrepancy between reproductive and social maturity (Gluckman and Hanson 2006a). Teen mothers are increasingly less prepared to become caretakers given the realities and complexities of growing up in a skill- and education-based economy where women often live apart from or cannot count on familial support.

These social conditions shape public attitudes and health policies about teen motherhood. Is a young mother able to care for a child's postnatal needs on her own? If not, can she count on support from kin-based childrearing networks? What are the opportunity costs and lifetime consequences of an early age at first birth? The answers to these questions are of principal concern in developed societies, where support for teen mothers tends to take the form of government-funded public assistance programs, often leading to years of protracted economic dependency and a cascade of social repercussions not paralleled in traditional societies. Government agencies stand as a poor substitute for the quality of caretaking bestowed upon a young mother and child by traditional kin-based support networks, which sustain successful childrearing and development. Public health research that further evaluates the biological consequences of teen motherhood, therefore, may be less valuable than studies that focus primarily on the various social costs and benefits of an early age at first birth.

5.8   Conclusion

Both biological and sociocultural factors affect the costs and benefits of teen motherhood. Except for the youngest of teens, who are clearly at a disadvantage, the causes and consequences of teen motherhood are socioecologically dependent. Because age-related biological and social costs are not universal, the decision to delay or initiate childbearing varies with cultural context and an individual's social and economic environment. Three points can be made about teen motherhood from the case study presented here on a traditional group of hunter-gatherers. First, the Pumé analysis concurs with the reproductive biology literature that the youngest of teen mothers (under 14.5 years) have an elevated risk of losing their first born. Second, from the perspective of a young woman starting her reproductive career, the optimal strategy to minimize infant deaths is to wait until her mid-teens to have her first child. Third, a young woman gains no additional fitness advantage, in terms of lowering infant mortality or raising lifetime fertility, by delaying childbearing until her late teens.

While the biological constraints and selective pressures on initiating first birth too young are expected to be similar across human populations, sociocultural and economic factors are highly variable. Many of the adverse effects associated with teen motherhood in developed societies are mediated through strong social support networks in traditional societies. However, young women growing up in developed societies may face different trade-offs because they often cannot count on the same level of social support in rearing children. Importantly, the decision to initiate childbearing is weighted against opportunities for education, training, and employment. A life history perspective helps to explain why teen motherhood generates considerable public, media, and scholarly concern in developed societies, yet is the norm in traditional societies, as it likely was in the ancestral past.

References

Adair L (2001) Size at birth predicts age at menarche. Pediatrics 107:59–65

Akinbami LJ, Schoendorf KC, Kiely JL (2000) Risk of preterm birth in multiparous teenagers. Arch Pediatr Adolesc Med 154:1101–1107

Barreto D, Rivas PJ (2007) Los Pumé (Yaruro). In: Freire G, Tillett A (eds) Salud indígena en Venezuela, vol 2. Caracas: gobierno bolivariano de Venezuela, Ministerio de Poder Popular para la Salud, Editorial Arte, p 247

Belsky J, Steinbery L, Draper P (1991) Childhood experience, interpersonal development, and reproductive strategy: an evolutionary theory of socialization. Child Dev 62:647–670

Bogaert AF (2005) Age at puberty and father absence in a national probability sample. J Adolesc 28:541–546

Boyce WT, Ellis BJ (2005) Biological sensitivity to context: I. An evolutionary-developmental theory of the origins and functions of stress reactivity. Dev Psychopathol 17(2):271–301

Charnov EL (1992) Life history invariants: some explanations of symmetry in evolutionary ecology. Oxford University Press, Oxford.

Chen XK, Wen SW, Fleming N et al (2007a) Teenage pregnancy and congenital anomalies: which system is vulnerable? Hum Reprod 22(6):1730–1735

Chen X, Wen SW, Fleming N, Demissie K, Rhoads GG, Walker M (2007b) Teenage pregnancy and adverse birth outcomes: a large population based retrospective cohort study. Int J Epidemiol 36:368–373

Chen XK, Wen SW, Fleming N et al (2008) Increased risks of neonatal and postneonatal mortality associated with teenage pregnancy had different explanations. J Clin Epidemiol 61:688–694

Chisholm JS (1999) Death, hope and sex: steps to an evolutionary ecology of mind and mortality. Cambridge University Press, New York

Chisholm JS, Quinlivan JA, Petersen RW, Coall DA (2005) Early stress predicts age at menarche and first birth, adult attachment, and expected lifespan. Hum Nat 16(3):233–265

Clutton-Brock TH (1988) Reproductive success. University of Chicago Press, Chicago

Coall DA, Chisholm JS (2003) Evolutionary perspectives on pregnancy: maternal age at menarche and infant birth weight. Soc Sci Med 57:1771–1781

Conde-Agudelo A, Belizán JM, Lammers C (2005) Maternal-perinatal morbidity and mortality associated with adolescent pregnancy in Latin America: Cross-sectional study. American Journal of Obstetrics and Gynecology 192:342–349

Cooper C, Kuh D, Egger P, Wadsworth M, Barker D (1996) Childhood growth and age at menarche. Br J Obstet Gynaecol 103:814–817
Currie C, Ahluwalia N, Godeau E, Gabhainn S, Due P, Currie DB (2012) Is obesity at individual
and national level associated with lower age at menarche? Evidence from 34 countries in the
health behaviour in school-aged children study. J Adolesc Health 50:621–626
Debiec KE, Paul KJ, Mitchell CM et al (2010) Inadequate prenatal care and risk of preterm deliv-
ery among adolescents: a retrospective study over 10 years. Am J Obstet Gynecol 203:122.
e1–122.e6
Early J, Headland TN (1998) Population dynamics of a Philippine rain forest people. University of
Florida Press, Gainesville
Early JD, Peters JF (1990) The population dynamics of the Mucajai Yanomama. Academic Press,
New York
Eder JF (1987) On the road to tribal extinction: depopulation, deculturation and adaptive well
being among the Batak of the Philippines. University of California Press, Berkeley
Ellis BJ (2004) Timing of pubertal maturation in girls: an integrated life history approach. Psychol
Bull 130(6):920–958
Ellis BJ, Bates JE, Dodge KA, Fergusson DM, Horwood LJ, Pettit GS, Woodward L (2003) Does
father absence place daughters at special risk for early sexual activity and teenage pregnancy?
Child Dev 74(3):801–821
Ellis BJ, Figueredo AJ, Brumbach BH, Schlomer GL (2009) Fundamental dimensions of environ-
mental risk: the impact of harsh versus unpredictable environments on the evolution and devel-
opment of life history strategies. Hum Nat 20:204–268
Ellison PT (1981) Prediction of age at menarche from annual height increments. Am J Phys
Anthropol 56:71–75
Ellison PT (1990) Human ovarian function and reproductive ecology: new hypotheses. Am
Anthropol 92(4):933–952
Ellison PT (2001) On fertile ground: a natural history of human reproduction. Harvard University
Press, Cambridge, MA
Ernande B, Dieckmann U, Heino M (2004) Adaptive changes in harvested populations: plasticity
and evolution of age and size at maturation. Proc R Soc Lond B 271:415–423
Eveleth PB, Tanner JM (1990) Worldwide variation in human growth. Cambridge University
Press, Cambridge
Forrest JD (1993) Timing of reproductive life stages. Obstet Gynecol 82:105–111
Foster A, Menken J, Chowdhury AI, Trussell J (1986) Female reproductive development: a haz-
ards model analysis. Soc Biol 33:183–198
Fraser AM, Brockert JE, Ward RH (1995) Association of young maternal age with adverse repro-
ductive outcomes. N Engl J Med 332(17):1113–1117
Gadgil M, Bossert WH (1970) Life historical consequences of natural selection. Am Nat
104(935):1–24
Garn SM (1987) The secular trend in size and maturational timing and its implications for nutri-
tional assessments: a critical review. J Nutr 117:817–823
Geronimus AT (1986) The effects of race, residence, and prenatal care on the relationship of mater-
nal age to neonatal mortality. Am J Public Health 76:1416–1421
Geronimus AT (1987) On teenage childbearing and neonatal mortality in the United States. Popul
Dev Rev 13(2):245–279
Geronimus AT (1996a) What teen mothers know. Hum Nat 7:323–352
Geronimus AT (1996b) Black/white differences in the relationship of maternal age to birthweight:
a population-based test of the weathering hypothesis. Soc Sci Med 42(4):589–597
Geronimus AT (2003) Damned if you do: culture, identity, privilege, and teenage childbearing in
the United States. Soc Sci Med 57:881–893
Geronimus AT, Bound J, Waidmann TA (1999) Health inequality and population variation in
fertility-timing. Soc Sci Med 49:1623–1636
Gilbert W, Jandial D, Field N, Bigelow P, Danielsen B (2004) Birth outcomes in teenage pregnancies. J Matern Fetal Neonatal Med 16:265–270

Gillett Netting R, Meloy M, Campbell BC (2004) Catch-up reproductive maturation in rural Tonga girls, Zambia? Am J Hum Biol 16:658–669

Gluckman PD, Hanson MA (2006a) Changing times: the evolution of puberty. Mol Cell Endocrinol 254–255:26–31

Gluckman PD, Hanson MA (2006b) Evolution, development and timing of puberty. Trends Endocrinol Metab 17:7–12

Greaves RD, Kramer KL (2013) Hunter-gatherer use of wild plants and domesticates: archaeological implications for mixed economies before agricultural intensification. J Archaeol Sci 41:263–271

Guttmacker Institute (2014) http://www.guttmacher.org/pubs/USTPtrends10.pdf

Harrison KA (1985) Child-bearing, health and social priorities: a survey of 22,774 consecutive hospital births in Zaria, northern Nigeria. Br J Obstet Gynaecol 5:1–119

Hawkes K, O'Connell J, Blurton Jones N (1997) Hadza women's time allocation, offspring provisioning and the evolution of long postmenopausal life spans. Curr Anthropol 38(4):551–577

Hewlett B (1991) Demography and childcare in preindustrial societies. J Anthropol Res 47(1):1–37

Hill K, Hurtado AM (1996) Ache life history. Aldine de Gruyter, New York

Hill K, Kaplan H (1999) Life history traits in humans: theory and empirical studies. Annu Rev Anthropol 28:397–430

Holmes R (1995) Small is adaptive: nutritional anthropometry of native Amazonians. In: Sponsel L (ed) Indigenous peoples and the future of Amazonia: an ecological anthropology of an endangered world. University of Arizona Press, Tucson

Howell N (2000 [1979]) Demography of the Dobe !Kung. Academic Press, New York

Hulanicka B, Gronkiewicz L, Koniarek J (2001) Effect of familial distress on growth and maturation in girls: a longitudinal study. Am J Hum Biol 13:771–776

Hurtado AM, Hawkes K, Hill K, Kaplan H (1992) Trade-offs between female food acquisition and child care among Hiwi and Ache foragers. Hum Nat 3(3):1–28

Ibáñez L, Ferrer A, Marcos MV, Hierro FR, de Zegher F (2000) Early puberty: rapid progression and reduced final height in girls with low birth weight. Pediatrics 106(5):e72–e74

Jasienska G, Ellison PT (2004) Energetic factors and seasonal changes in ovarian function in women from rural Poland. Am J Hum Biol 16:563–580

Jolly MC, Sebire N, Harris J, Robinson S, Regan L (2000) Obstetric risks of pregnancy in women less than 18 years old. Obstet Gynecol 96:962–966

Jones ME, Cockburn A, Hamede R, Hawkins C, Hesterman H, Lachish S, Mann D, McCallum H, Pemberton D (2008) Life-history change in disease-ravaged Tasmanian devil populations. Proc Natl Acad Sci 105(29):10023–10027

Kassebaum NJ et al (2014) Global, regional and national levels and causes of maternal mortality during 1990-2013: a systematic analysis for the Global Burden of Disease Study 2013. Lancet 384:980–1004

Kim K, Smith PK, Palermiti A (1997) Conflict in childhood and reproductive development. Evol Hum Behav 18:109–142

King JC (2003) The risk of maternal nutritional depletion and poor outcomes increases in early or closely spaced pregnancies. J Nutr 133(5):1732S–1736S

Koniak-Griffin D, Turner-Pluta C (2001) Health risks and psychosocial outcomes of early childbearing: a review of the literature. J Perinat Neonatal Nurs 15:1–17

Koziel S, Jankowska EA (2002) Effect of low versus normal birthweight on menarche in 14-year old Polish girls. J Paediatr Child Health 38(3):268–271

Kramer KL (2005) Maya children: helpers at the farm. Harvard University Press, Cambridge, MA

Kramer KL (2008) Early sexual maturity among Pumé foragers of Venezuela. Fitness implications of teen motherhood. Am J Phys Anthropol 136(3):338–350

Kramer KL (2010) Cooperative breeding and its significance to the demographic success of humans. Annu Rev Anthropol 39:414–436

Kramer KL (2011) The spoken and unspoken. In: Canfield MR (ed) Field notes for scientists and naturalists. Harvard University Press, Cambridge, MA

Kramer KL, Ellison PT (2010) Pooled energy budgets: resituating human energy allocation trad-eoffs. Evol Anthropol 19:136–147

Kramer KL, Greaves RD (2007) Changing patterns of infant mortality and fertility among Pumé foragers and horticulturalists. Am Anthropol 109(4):713–726

Kramer KL, Greaves RD (2011) Juvenile subsistence effort, activity levels and growth patterns: middle childhood among Pumé foragers. Hum Nat 22:303–326

Kramer M, Demissie K, Yang H, Platt R, Sauve R, Liston R (2000) The contribution of mild and moderate preterm birth to infant mortality. JAMA 284(7):843–849

Kramer KL, Greaves RD, Ellison PT (2009) Early reproductive maturity among Pumé foragers. Implications of a pooled energy model to fast life histories. Am J Hum Biol 21(4):430–437

Kramer KL, Lancaster JB (2010) Teen motherhood in cross-cultural perspective. Annals of Human Biology 37(5):613–628

Kuzawa CW (2005) Fetal origins of developmental plasticity: are fetal cues reliable predictors of future nutritional environments? Am J Hum Biol 17:5–21

Kuzawa CW, Sweet E (2009) Epigenetics and the embodiment of race: developmental origins of US racial disparities in cardiovascular health. Am J Hum Biol 21:2–15

Lancaster JB, Hamburg BA (eds) (2008 [1986]) School-age pregnancy and parenthood. Aldine de Gruyter and Transaction Publishers, New York and Piscataway, NJ

Lancaster JB, Kaplan H, Hill K, Hurtado AM (2000) The evolution of life history, intelligence and diet among chimpanzees and human foragers. In: Tonneau F, Thompson NS (eds) Perspectives in ethology: evolution, culture and behavior. Kluwer, New York

Lee KS, Ferguson RM, Corpuz M, Gartner LM (1988) Maternal age and incidence of low birth weight at term: a population study. Am J Obstet Gynecol 158(1):84–89

Mahavarkar SH, Madhu CK, Mule VD (2008) A comparative study of teenage pregnancy. J Obstet Gynaecol 28(6):604–607

Makinson C (1985) The health consequences of teenage fertility. Fam Plann Perspect 17(3):132–139

Meade CS, Kershaw TS, Ickovics JR (2008) The intergenerational cycle of teenage motherhood: an ecological approach. Health Psychol 27(4):419–429

Migliano AB, Vinicius L, Lahr MM (2007) Life history trade-offs explain the evolution of human pygmies. Proc Natl Acad Sci 104(51):20216–20219

Morris DH, Jones ME, Schoemaker MJ, Ashworth A, Swerdlow AJ (2011) Secular trends in age at menarche in women in the UK born 1908–93: results from the Breakthrough Generations Study. Paediatr Perinat Epidemiol 25:394–400

Nichols HB, Trentham-Dietz A, Hampton JM, Titus-Ernstoff L, Egan KM, Willett WC et al (2006) From menarche to menopause: trends among US women born from 1912 to 1969. Am J Epidemiol 164:1003–1011

Olausson PO, Cnattingius S, Haglund B (1999) Teenage pregnancies and risk of late fetal death and infant mortality. Br J Obstet Gynaecol 106(2):116–121

Olsen EM, Heino M, Lilly GR, Morgan MJ, Brattey J, Ernande B, Dieckmann U (2004) Maturation trends indicative of rapid evolution preceded the collapse of northern cod. Nature 428(29):932–935

Parent AS, Teilmann G, Juul A, Skakkebaek NE, Toppari J, Bourguignon JP (2003) The timing of normal puberty and the age limits of sexual precocity: variations around the world, secular trends, and changes after migration. Endocr Rev 24:668–693

Pennington R (2001) Hunter-gatherer demography. In: Panter-Brick C, Layton RH, Rowley-Conwy P (eds) Hunter-gatherers: an interdisciplinary approach. Cambridge University Press, Cambridge

Perry JRB, Day F, Elks CE et al (2014) Parent-of-origin-specific allelic associations among 106 genomic loci for age at menarche. Nature 514:92–97

Phipps MG, Sowers M (2002) Defining early adolescent childbearing. Am J Public Health 92(1):125–128

Phipps MG, Blume JD, DeMonner SM (2002) Young maternal age associated with increased risk of postneonatal death. Obstet Gynecol 100:481–486

Population Reference Bureau (2007) http://www.prb.org/datafind/prjprbdata. Accessed 15 July 2007

Preston SH, Heuveline P, Guillot M (2001) Demography: measuring and modeling population processes. Blackwell, Oxford

Quinlan RJ (2003) Father absence, parental care and female reproductive development. Evol Hum Behav 24:376–390

Riley AP, Samuelson JL, Huffman SL (1993) The relationship of age at menarche and fertility in undernourished adolescents. In: Gray RH, Leridon H, Spira A (eds) Biomedical and demographic determinants of reproduction. Clarendon, Oxford

Salzano FM, Callegari-Jacques SM (1988) South American Indians: a case study in evolution. Clarendon, Oxford

Satin AJ, Leveno KJ, Sherman ML, Reedy NJ, Lowe TW, McIntire DD (1994) Maternal young and pregnancy outcomes: middle school versus high school age groups compared to women beyond their teen years. Am J Obstet Gynecol 171:184–187

Sellen D (2000) Seasonal ecology and nutritional status of women and children in a Tanzania pastoral community. Am J Hum Biol 12:758–781

Sellen D (2006) Lactation, complementary feeding, and human life history. In: Hawkes K, Paine RR (eds) The evolution of human life history. School of American Research Press, Santa Fe

Stearns SC (1989) Trade-offs in life-history evolution. Funct Ecol 3:259–268

Stevens-Simon C, McAnarney ER (1995) Further evidence of reproductive immaturity among gynecologically young pregnant adolescents. Fertil Steril 64:1109–1112

Stevens-Simon C, Beach RK, McGregor JA (2002) Does incomplete growth and development predispose teenagers to preterm delivery: a template for research. J Perinatol 22(4):315–323

Towne B, Czerwinski SA, Demerath EW, Blangero J, Roche AF, Siervogel RM (2005) Heritability of age at menarche in girls from the Fels longitudinal study. Am J Phys Anthropol 128:210–219

U.S. Department of Health and Human Services (2015) Office of Adolescent Health. http://www.hhs.gov/ash/oah/adolescent-health-topics/reproductive-health/teen-pregnancy/trends.html. Accessed 23 Jan 2015

Ulijaszek S (1995) Human energetics in biological anthropology. Cambridge University Press, Cambridge

United Nations Economic Commission for Europe (2007) www.unece.org. Accessed 15 July 2007

Valeggia CR, Ellison PT (2004) Lactational amenorrhea in well nourished Toba women of Argentina. J Biosoc Sci 36:573–595

Walker R, Gurven M, Hill K, Migliano A, Chagnon N, De Souza R, Djurovic G, Hames R, Hurtado AM, Kaplan H et al (2006) Growth rates and life histories in twenty-two small-scale societies. Am J Hum Biol 18:295–311

Wood JW (1992) Fertility and reproductive biology in Papua New Guinea. In: Attenborough R, Alpers M (eds) Human biology in Papua New Guinea: the small cosmos. Clarendon, Oxford

Worthman CM (1993) Biocultural interactions in human development. In: Pereira ME, Fairbanks LA (eds) Juvenile primates. Oxford University Press, Oxford

Chapter 6
Health, Evolution, and Reproductive Strategies in Men: New Hypotheses and Directions

Richard G. Bribiescas and Erin E. Burke

Abstract This chapter presents new research directions and hypotheses based on the evolutionary relationships between reproductive effort and contemporary health challenges in men. Drawing on life history, these hypotheses are grounded in the observation that human male reproductive strategies exhibit a much broader range of variability compared to other great apes. Specifically, human males are unique in that they exhibit the capacity to devote a significant amount of time and energy to offspring and mate care, often spending much more of their lives providing paternal investment compared to other primates and indeed many mammals. Men also exhibit the capability to negotiate and partition investment in offspring according to perceived paternity. The extraordinary range of human male reproductive options can be viewed as selection for neuroendocrine adjustments in response to shifting mortality challenges and extended life spans. We hypothesize that (1) investment in offspring and mates evolved in tandem with decreases in human male mortality and morbidity, (2) male fertility at older ages coevolved with an increased capacity to defend against degenerative diseases, (3) testosterone and other neuroendocrine mechanisms are primary targets of selection for the evolution of these traits, and (4) trade-offs between male health maintenance, variation in paternal investment, and male reproductive strategies are contingent on energetic constraints. Drawing on research from the literature, we provide a number of case studies that inform our hypotheses and provide guidance for future research.

R.G. Bribiescas (✉) • E.E. Burke
Department of Anthropology, Yale University,
10 Sachem Street, New Haven, CT 06520, USA
e-mail: richard.bribiescas@yale.edu

© Springer Science+Business Media New York 2017
G. Jasienska et al. (eds.), *The Arc of Life*, DOI 10.1007/978-1-4939-4038-7_6

6.1 Introduction: Health and Male Reproduction

The males of species in which mating opportunities produce a broad range of fitness variation are likely to exhibit higher mortality and morbidity compared to females. If the fitness benefits of reproductive behavior outweigh the survivorship costs, selection will still favor it. Therefore, it is not surprising that mammals such as men often exhibit higher age-specific mortality, shorter life spans, and greater morbidity compared to women (Austad 2006, 2011). The reasons behind this sex disparity have been extensively debated among clinicians and epidemiologists. However, sex differences in human mortality and longevity come as no surprise to evolutionary biologists: they not only occur in many nonhuman organisms but are predicted by evolutionary theory (Andersson 1994).

We suggest that mortality and morbidity variation are entwined with variation in human male reproductive strategies. That is, as male mortality and morbidity challenges vary, so will male reproductive biology and vice versa. In essence, there is a tension between traits that are beneficial to reproductive success early in life and downstream costs to well-being and life span that have likely been important to the evolution of our genus (Williams 1957; Kirkwood and Austad 2000). This antagonistic pleiotropy is at the root of much of the senescence theory. It occurs when a single gene has multiple effects on an organism. Antagonist pleiotropy refers to a situation where there is selection for a trait with beneficial early effects and deleterious later effects. Antagonistic pleiotropy was first implicated in theories of senescence by Medawar and further developed by Williams (Williams 1957; Medawar 1952). They recognized that traits that are beneficial to reproduction early in life but have negative health consequences later in life will be maintained in the gene pool because they increase the early-life fitness of the carrier. These late-life negative health consequences are not selected against and thus lead to senescence (see Chap. 9 for a further discussion of senescence theories) (Kirkwood and Austad 2000; Partridge and Barton 1993; Finch and Rose 1995).

Discussions of health and male reproduction commonly converge on questions of male fertility. This chapter, however, aims to reframe this association by suggesting that human male health is broadly influenced by the evolutionary biology of reproduction and behaviors that are indicative of male reproductive effort. That is, to understand the health challenges of human males, it is necessary to comprehend the evolutionary and pleiotropic relationships between morbidity, mortality, reproductive effort, and its underlying physiology. It would be constructive to view men through the lens of life history theory.

In a 2006 *New York Times* article, Dr. Demetrius J. Porche (then an associate dean at Louisiana State University's Health Sciences Center School of Nursing and current editor in chief of the *American Journal of Men's Health*) stated:

> "We keep throwing out lifestyle as an explanation for the differences in longevity, saying that men come in later for care and have unhealthy behaviors, but I'm not sure we really know the reason." Dr. Porche went on to say, "And we haven't answered the question: Is there a biological determinant for why men die earlier than women?" (Rabin 2006)

The short answer is "yes." Regardless of population, culture, or environmental variability, human male survivorship at all life stages is often compromised compared to females. However, in many economically developing countries, women have higher mortality during their childbearing years due to maternal mortality. This sex disparity extends to comparative studies of other primates, including our closest evolutionary relative, the chimpanzee (Bronikowski et al. 2011; Hill et al. 2001). Differences are often evident even in utero and the juvenile period (childhood) but they are most prominent in adulthood (Wells 2000).

Differences in life span between men and women are likely due to a combination of compromised survivorship in men and enhanced survivorship in women (Schroder et al. 1998). Compared to women, men are often more vulnerable to infection and have poorer prognoses although it is important to remember that women also exhibit sex-specific vulnerabilities such as autoimmune disorders (Austad 2006). Physicians and clinical researchers have struggled with this question since the relationship between being male and getting sick is robust. Within the clinical community, there is also the misguided assumption that millions of years of evolution would result in a physiology that optimizes health and survivorship. But this runs contrary to evolutionary thinking. Natural selection shapes physiology to optimize lifetime evolutionary fitness, but does not necessarily promote physical or behavioral characteristics that are commonly associated with health. When somatic maintenance converges with reproductive effort, selection will favor traits that make us healthy and happy. However, evolution may not favor this convergence, particularly in men since reproductive effort often increases at the expense of morbidity and survivorship. In essence, overall health in men and women is inextricably entwined with reproductive trade-offs (Ellison and Jasienska 2009).

In this chapter we discuss several aspects of health related to the male life history, beginning with mortality during the transition to reproductive maturation, and then moving on to discuss the broad range of male reproductive strategies, health implications, selected major health issues, and the neuroendocrine factors that often guide the phenotypic expression and plasticity of aspects of male physiology that underlie health challenges. Concluding, we discuss future directions whereby human evolutionary biology can inform our understanding of men's health.

6.2 Male Health and Reproductive Maturation

The transition from childhood to reproductive maturity in males involves the redirection of energetic and behavioral investment from general somatic growth and development to physiological functions related more directly to reproductive effort. This includes the growth of sexually dimorphic tissues such as bone and skeletal muscle, spermatogenesis, and an increased motivation toward mate-seeking behavior. This transition is also associated in all human societies—traditional and modern with a spike in mortality in young males that is correlated with risky behaviors (Fig. 6.1).

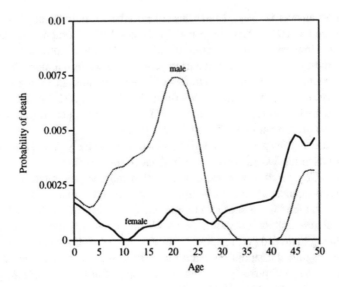

Fig. 6.1 Sex differences in mortality in Aché foragers of Paraguay (Hill and Hurtado 1996)

Risk-taking behavior during adolescence and young adulthood is an important public health concern worldwide. Many young male deaths occur as a consequence of accidents, homicides, suicides, substance abuse, war, and other risky or violent behaviors (Daly and Wilson 1983a). We categorize these causes of mortality as "extrinsic mortality" in contrast to intrinsic causes of mortality (those that lead to death via senescence or disease) (Partridge and Barton 1993; Carnes et al. 2006). Males, especially young adult males, are much more likely to engage in these activities compared to females than (Byrnes et al. 1999). For example, Wang and colleagues (2009) found that male college students in the USA are more likely to participate in risky behavior of all types, from chasing a bear out of a campsite to having unprotected sex. Males also commit more violent crimes, which can be seen in the higher rates of men killing unrelated men in Chicago, Detroit, England, and Canada (Daly and Wilson 1983b). These examples provide a snapshot of the ways in which males in industrialized societies engage in risky behavior; comparable scenarios can be found in all human cultures. In fact, this discrepancy between male and female risk taking and subsequent mortality follows the general mammalian pattern. In order to understand the ultimate cause of this discrepancy, it is important to consider sexual selection theory in general.

Greater male extrinsic mortality compared to females is central to understanding sex-specific reproductive strategies (Bonduriansky et al. 2008; Trivers 1972). Females tend to engage in significant metabolic investment in reproduction compared to males. The production of energetically expensive, internally fertilized eggs, gestation, lactation, offspring care, and provisioning compared to the relatively low metabolic costs associated with spermatogenesis underlie the contrasting energetic investments between men and women. Female fitness is most often limited by access to resources to support the energetic costs of reproduction. Males, on the other hand, are largely unencumbered by the direct energetic requirements of reproduction related to gamete production, gestation, and lactation. Male fitness is then not limited by resources, but by access to females and energetic investment in somatic tissue that is commonly deployed toward reproductive effort, such as sexually dimorphic muscle tissue (Bribiescas 1996).

This difference in reproductive investment has important implications for the variance of reproductive success between males and females. A male can potentially impregnate many females at the same time and conceive many offspring. Females, however, have much less variance in their fitness because they have a finite number of resources and time in which to reproduce. A commonly cited figure for most offspring sired by a single human male is 888, although historical records report approximately 1157 (Oberzaucher and Grammer 2014; Busnot 1712). In contrast, the current record holder for the most offspring born to a female without reproductive technology is 69 offspring (Clay 1989). Between-male variance in fitness is also greater. In those societies where the level of effective polygyny and male fitness variation is high, risk taking tends to be elevated (Kruger and Nesse 2006). This idea also holds when considering economic inequality, as access to resources and status confer fitness advantages to males.

Differences in potential evolutionary payoffs are predicted to lead to differences in male and female mating strategies and behavior (Bonduriansky et al. 2008; Vinogradov 1998; Bateman 1948). When mating opportunities are limited, males compete for mating opportunities with other males because the differential in reproductive payoffs from engaging in this competition can potentially be great. Consequently, the payoffs vary more in men compared to women in polygynous societies, although humans exhibit a greater range of variation in mating strategies and behaviors compared to other primates (Brown et al. 2009). Also see Kokko and Jennions (2008). The timing of this risk taking during the life history is also important. It pays to take the greatest risks at the beginning of one's reproductive career, when the mating "market" is more open because females have not yet chosen long-term mates. The adolescent period can thus be viewed as an "inflection point in developmental trajectories of status, resource control, mating success, and other fitness-relevant outcomes" (Ellis et al. 2012, p. 601). It is consistent with this view, then, that male adolescents and young adults experience the highest rates of mortality. The male/female mortality ratio peaks at 3.01 in the 20–24-year age class in US males in 2000, and if one only considers extrinsic mortality (deaths from causes unrelated to illness), that ratio increases to 4.03 (Kruger and Nesse 2006).

However as anthropologists, we are often interested in the applicability of modern circumstances to the conditions that were more common during human evolution. In essence, our current environment of sedentism and high caloric availability is likely very different from our evolutionary past. It is therefore common for anthropologists to look at hunter-gatherer populations for an alternative perspective. This research strategy is not comprehensive as forager populations vary considerably in their ecologies and environments. However, it does provide a useful socioecological model to compare with modern industrial societies. Among the Aché of eastern Paraguay, the male/female mortality ratio is 1.8 for this same age class (Hill and Hurtado 1996). The difference in the mortality ratios between the Aché, a traditional society, and industrial societies, such as the USA, is not surprising since the discrepancy between male and female mortality rates has been increasing in developed nations for the last century. Kruger and Nesse attribute this to an overall decrease in infectious disease in industrialized societies, leaving a larger portion of mortality risks attributable to risky behavior undertaken primarily by men (Kruger and Nesse 2006). Male-skewed mortality during adolescence has been viewed as a public health problem, so it would be useful to deploy a life history perspective to address this challenge (Bell et al. 2013).

Recently, several authors have drawn attention to this discrepancy in mortality between the sexes. Ellis and colleagues (2012) make a persuasive case for the adoption by policymakers, scientists, and lawmakers of what they term the "evolutionary model" of adolescent risk taking, as opposed to the "developmental psychopathology model." In the developmental psychopathology model, risk taking is considered maladaptive because it threatens a person's "health, development, and safety," while the evolutionary model focuses on the potential fitness benefits of such risky behavior (Ellis et al. 2012). Kruger has also highlighted the need for public health professionals to embrace life history theory when confronted with risk-taking behaviors and strategies. He emphasizes the need for public health officials and researchers to be aware of life history strategies in humans that manifest themselves in earlier reproduction and higher homicide rates, for example (Kruger 2011).

Moving beyond a basic understanding of the evolutionary forces behind risk taking and other competitive and violent behaviors that lead to higher male mortality, how can life history theory inform interventions aimed at curbing male mortality? If risky behavior has potential fitness benefits, this could pose a challenge for those attempting to develop interventions aimed at curbing male mortality. By understanding the motives of males who undertake risks, namely, increases in status, one can then meet those goals in less lethal ways. Ellis and colleagues cite an example wherein schoolchildren earn points for the group by refraining from exhibiting "problem behaviors" (Ellis et al. 2012).

Kruger, on the other hand, emphasizes the reduction in status differentials between males by using egalitarian societies as a possible model (Kruger 2011). Regardless of the methods proposed, it is exciting that researchers are exploring interventions that consider the evolutionary rationale of such behavior. Hopefully, future health policy will incorporate an understanding of sexual selection and its effects on male risk taking when implementing measures to curb male mortality.

While peak lifetime testosterone levels coincide with risky behavior and the spike in male mortality in the late teen years and early twenties, the actual relationship between circulating testosterone and young male risk taking is less straightforward. While testosterone is highest during the second decade of life in many economically developed populations, foragers and males under greater energetic stress exhibit much smaller or no age-related differences in testosterone levels (Ellison et al. 2002; Uchida et al. 2006). Therefore, it is likely that pubertal testosterone increases are likely to be more contributory to risky behavior and mortality regardless of absolute adult-level testosterone. Additional increases in testosterone in more energetically rich populations probably reflect priming of luteinizing hormone (LH) and gonadotropin-releasing hormone (GnRH) receptors and greater somatic investment in lean body mass (Smith et al. 1975; Spratt and Crowley 1988).

Overall, adolescence is an important life stage in men, one in which mortality rates are significantly greater than in females. Comparative investigations strongly suggest that this period of high male mortality during reproductive maturation is a conserved trait common to many mammals, including primates (Pereira and Fairbanks 2002). Tolerance for risk and a hampered ability to accurately assess hazards contribute to health threats in young men (Cohn et al. 1995). Looking beyond testosterone, serotonin levels are associated with greater mortality in young male primates (*Macaca mulatta*), possibly hindering risk assessment and increasing risk tolerance (Higley et al. 1996).

We hypothesize that the development of interventions to decrease young male mortality is challenging given the potential fitness benefits that could promote risky behavior, as well as the possibility of transgenerational influences that may contribute to population differences in morbidity and mortality (Jasienska 2009). Nonetheless, changes in perceptions of environmental risk and socioeconomic factors hold some promise for promoting positive change. For example, van Anders et al. (2012) reported that testosterone was positively associated with safer sex attitudes, especially those most closely tied to STI (sexually transmitted infection) risk avoidance. Among some African American communities, young male mortality remains stubbornly high compared to that in the general population. An examination of 66 US counties that exhibit African American mortality rates in line with those in the general population report, not surprisingly, that income, poverty, and education are delineating factors between counties with high and low African American mortality (Levine et al. 2013). These factors in turn strongly contribute to perceptions of future survivorship and well-being.

6.3   Health and Fatherhood

If male mortality is high, evolutionary theory suggests that males should invest more in mating opportunities with multiple females, since this would provide the greatest increase in fitness in the face of shortened life spans (Trivers 1972). The evolution of significant paternal investment, on the other hand, should coincide with an increase in male survivorship. Moreover, paternal investment should be associated with an increase in offspring survivorship, although fathers can also place burdens on families (Hewlett 1992; Flinn and England 2003). But can fatherhood itself serve as a means of decreasing male morbidity and mortality?

Garfield et al. (2006) contend that fatherhood is indeed an important factor for understanding patterns of male morbidity and mortality. Among other observations, they suggest that fatherhood may improve men's health, since the responsibility of

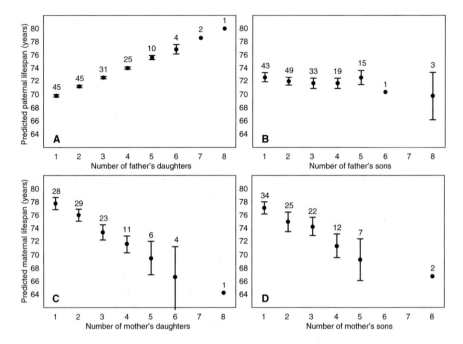

Fig. 6.2 A number of daughters are associated with longer paternal life spans among a rural Polish community. Sons had no effect. Numbers of both sons and daughters are associated with shorter postmenopausal maternal life span (Jasienska et al. 2006)

raising a child creates a greater awareness of the need to be mindful of a parent's own health and well-being. Fatherhood is also associated with lower health risks in some cultures. American men over the age of 50 with two or more children have a significantly lower risk of cardiovascular death compared to married men with no children (Eisenberg et al. 2011). A demographic study of rural Polish fathers reported that number of daughters significantly increased paternal survivorship, while number of sons had no effect (Fig. 6.2) (Jasienska et al. 2006). These researchers suggest that the patriarchal structure of rural Poland encourages daughters to stay home, contributing to household tasks, hygiene, and supporting their fathers' overall well-being, while sons are more likely to disperse from their homes.

In iteroparous species with altricial offspring (species that reproduce repeatedly and whose young mature more slowly), it is often beneficial for mothers to survive long enough to care for their offspring to mature. For males, this is often not the case. Internal fertilization results in uncertainty of paternity, which is predicted to compromise the potential for male paternal investment. Therefore, instead of investing in paternal care and long lives (like females), males benefit by investing in male-male competition for access to fertile females. While mating with multiple males also may augment female fitness in a number of ways, greater offspring output is not one of them (Hrdy 2000). This pattern holds for most mammals; however, humans exhibit a unique divergence from the common mammalian pattern. Humans

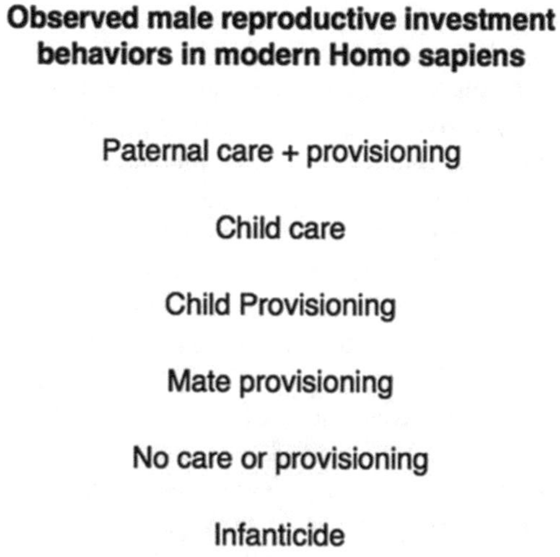

Fig. 6.3 Range of human male offspring investment (Bribiescas et al. 2012)

Observed male reproductive investment behaviors in modern Homo sapiens

Paternal care + provisioning

Child care

Child Provisioning

Mate provisioning

No care or provisioning

Infanticide

have diverged from the typical mammalian male mating schema, in which fathers contribute little to offspring care; we are among the few primates, and only great ape species, that engages in significant care of both young and mates (Gray and Anderson 2010; Bribiescas et al. 2012; Gettler 2010). Moreover, human males have evolved the broadest range of paternal investment patterns of any primate, including extensive paternal care and provisioning (Fig. 6.3).

The extensive (and presumably adaptive) range of behavioral reproductive strategies in men has required the evolution of the neuroendocrine plasticity that is conducive to paternal offspring investment and pair-bonding. To date, numerous studies have reported that testosterone levels are lower in fathers (Gray and Anderson 2010; Kuzawa et al. 2009; Gray 2003; Gray et al. 2002, 2007, 2006; Burnham et al. 2003), supporting the idea that there is adaptive phenotypic plasticity in these levels. In a longitudinal study, Gettler et al. (2011) noted a decline in testosterone in Filipino men after they became fathers, although they also report that young men who had attenuated declines in testosterone with fatherhood engaged in more sexual intercourse with their partners, compared to men who had significantly greater declines with fatherhood (Gettler et al. 2013). These data are consistent with the idea that environmental and social modulation of neuroendocrine function are reflected in the plasticity of male reproductive strategies.

The ability of organisms to respond adaptively to morbidity and mortality challenges is a major predictor of population survival, and a major tenet of life history theory is that natural selection in the form of mortality pressure will shape patterns of investment by males in offspring and mates. Hence a critical prediction of evolutionary theory is that male neuroendocrine function should respond adaptively to environmental challenges that correlate with mortality and morbidity risk. While the paternal neuroendocrine profile is complex and involves many hormones, such as oxytocin, vasopressin, prolactin, and testosterone, the latter is particularly important, since it is closely associated with variation in immunocompetence, morbidity, mortality, and risk tolerance.

Changes in testosterone levels in response to fatherhood may be an adaptive response to facilitate care and bonding behavior and decrease mate-seeking behavior. However, other possibilities include lowering testosterone to augment paternal immunocompetence (see Chap. 7). Testosterone lowers immune function, so men with high testosterone levels could run the risk of endangering their immunologically naïve offspring by transmitting contagious diseases (Muehlenbein and Bribiescas 2005). Moreover, decreases in testosterone may also facilitate the deposition of adipose tissue to aid in paternal survivorship during periods of energetic stress (Lakshman et al. 2010; Bribiescas 2001).

Several investigations have explored the relationship between health-seeking behavior and fatherhood, particularly in reference to the possibility of *couvade* or "sympathy pregnancy" symptoms presented by expectant fathers in association with the pregnancies of their partners (i.e., weight gain, nausea). Interestingly, American men with expectant wives visited health-care providers less during the 9 months of pregnancy compared to a control group of men without expectant wives. The decrease in health-seeking behavior during pregnancy did not support the prediction of couvade behavior; however, this may be due to the increased attention to the pregnant partner and her health needs (Fig. 6.4) (Quill et al. 1984).

We hypothesize that the evolution of paternal behavior coincided with decreases in morbidity and mortality in men, possibly in association with the emergence of the genus *Homo* (Bribiescas et al. 2012). Moreover, we suggest that the broad range of male offspring investment strategies now seen in human cultures was affected by males' ability to accurately assess and manage mortality risks. An intriguing example can be seen among indigenous Amazonian human groups, in which paternity is visibly parceled out and negotiated between men, resulting in the identification of

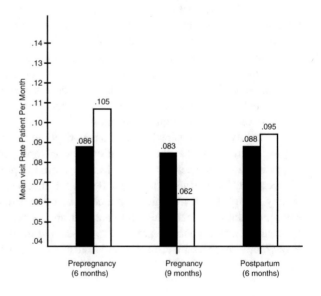

Fig. 6.4 Expectant fathers' health-care seeking immediately before and after the pregnancies of their partners. *Dark bars* are matched controls; *open bars* are expectant/ subsequent fathers (Quill et al. 1984)

primary, secondary, and sometimes tertiary fathers of a single offspring. This is likely a way to decrease social tensions, to increase ties with kin, and to manage conflict and male-male competition (Beckerman and Valentine, 2002; Walker et al. 2010). It is also likely that changes in hormone levels, particularly testosterone, in association with fatherhood, serve as an adaptive response to improve paternal survivorship and morbidity through the attenuation of the behavioral, immunological, and metabolic liabilities associated with testosterone (Gray and Anderson 2010).

6.4   Male Reproductive Effort and Cancer

Prostate cancer is a leading cause of death in men living in industrialized countries. Although there is a strong genetic component to prostate cancer risk (Amundadottir et al. 2006), an equally important environmental effect contributes to the risk of onset. Prostate cancer is exacerbated and possibly caused by chronic exposure to the sex steroids testosterone and dihydrotestosterone (DHT). Men in industrialized societies have higher overall levels of testosterone and live longer. Combined, this leads to a situation wherein men are exposed to more androgens over a longer period of time, increasing lifetime exposure rates and rates of emergent cancers that might not otherwise have been a cause of death. While the evidence of a direct dose-dependent effect on prostate cancer risk within individuals is conflicting, it is likely that chronic exposure to higher levels of testosterone and DHT may contribute to risk later in life. In essence, this cancer may result from an antagonistic pleiotropic effect, such that androgens incur a positive effect on fitness early in life, but a detrimental effect on survivorship at older ages at which reproductive potential has waned.

There is uneven evidence of an association between prostate cancer risk and within- or between-population differences in adult testosterone levels. Any association is apt to be rendered more complex by several factors. First, hormone measurements in these studies are usually taken only once or within a limited period of time during adulthood or at older ages. Prostate cancer risk, however, is likely to be the result of lifelong exposure to androgens, rather than levels measured at a given age or life stage. In addition, there are layers of variation that are often unaccounted for in studies that assess androgen levels with prostate cancer risk. Among men living in industrialized societies, testosterone levels can vary between healthy individuals by a factor of ten (Bribiescas 2001).

Alvarado (Calistro Alvarado 2010) addressed this issue by comparing testosterone levels in young men and prostate cancer disparities between different ethnic groups within the same American populations. His analysis provided compelling evidence that lifelong exposure to higher testosterone levels does contribute to prostate cancer risk. His results are similar to those of Jasienska and Thune (Jasienska and Thune 2001) showing salivary progesterone levels to be positively correlated with breast cancer risk between populations when assessed using the same assay methods and within the same laboratory.

Reproductive cancers are inextricably associated with reproductive maturity and function in men. While testosterone is a poor indicator of male fertility, its secondary effects on somatic investment and behavior are vital for men's reproductive fitness. We hypothesize that prostate cancer and chronically high testosterone levels are an example of *antagonistic pleiotropy, with testosterone promoting greater investment in tissue that is reflective of reproductive effort (i.e., sexually dimorphic muscle) at younger ages with later increases in cancer risk as male reproductive value declines with age.* Since there is significant male fertility at older ages across many populations (Tuljapurkar et al. 2007), it will be challenging to eradicate the incidence of this health risk, given that it is associated with reproductive benefits throughout men's life history.

Moreover, male reproductive status and effort may also be associated with prostate cancer risk. Many studies suggest that fathers have a greater risk of prostate cancer risk compared to childless men. However, many of these studies have included childless men who often have reproductive disorders, including impaired androgen function (Wiren et al. 2013). Hence, in many of these investigations, it is difficult to disentangle the relationship between infertility and fatherhood. However, in an epidemiological study of Danish men, when childless men were excluded, there was a significant inverse linear relationship between prostate cancer risk and number of children fathered, resulting in a 5 % reduction in prostate cancer risk per additional child (Jorgensen et al. 2008). While this is fragile evidence to support the idea that lower testosterone levels among fathers is associated with decreased prostate cancer risk, it does support the hypothesis of an association between prostate cancer risk and men's reproductive status.

6.5   Aging, Health, and Reproductive Options

As men age, the force of mortality shifts from extrinsic to intrinsic sources. Hormone levels also change with age, although there is considerable between- and within-population variation in the degree of age-related decreases in testosterone (Ellison et al. 2002; Harman et al. 2001). Other hormones, such as gonadotropins, tend to exhibit less population variation in association with age although few non-Western populations have been examined (Bribiescas 2005). While decreases in male fertility and reproductive hormones are likely influenced by decreased selection pressure, aging can also influence male reproductive strategies. While women undergo menopause around the age of 50, and men exhibit a "decline in fertility", significant potential for reproduction remains until death in a broad range of populations (Tuljapurkar et al. 2007). The prospect of continued fitness even at later ages opens the door for selection for reproductive strategies that are conducive with the inevitable somatic and health changes that occur with aging and senescence (Bribiescas 2006a, b, 2010).

Unlike other great apes (Muller and Wrangham 2004; Emery Thompson et al. 2012), human males appear to have evolved the capacity to decouple physical vigor from reproductive fitness. For example, among Aché men, physical strength is greatest in the early twenties (Walker and Hill 2003). However hunting efficiency, a trait that has been associated with reproductive success (Kaplan and Hill 1985), is highest when men are in their forties (Walker et al. 2002). The implication is that compared to other great apes, human male reproductive strategies have the capacity for some adjustment to somatic aging, most likely through leveraging greater cognitive capacity, larger brains, and the evolution of extrasomatic assets such as tools and social complexity.

Nonetheless, men must adapt to the natural degenerative changes that occur with aging. Indeed, given the long lives of humans, the capacity for significant fertility at older ages, and the broad range of reproductive strategies, there should be selection for protective mechanisms that attenuate age-related sources of intrinsic mortality and morbidity. One primary source of intrinsic mortality that is theorized to be associated with aging is oxidative stress and damage. Toxic reactive oxygen species (ROS), such as superoxide molecules, are created during aerobic metabolism and can outpace the body's ability to absorb or disarm them. There is evidence that excess ROS damage cellular components, including DNA, and this damage is implicated in many aging-related illnesses (Finkel and Holbrook 2000). Other downstream toxins such as hydrogen peroxide can induce lipid peroxidation and damage cellular components and also contribute to aging.

Given the shorter life spans of males in many species, Alonso-Alvarez theorized that testosterone-driven increases in metabolic rate during prenatal development may contribute to overall male fragility compared to females (Alonso-Alvarez et al. 2007). Data from zebra finches has provided some preliminary support for this theory (Tobler and Sandell 2009; Tobler et al. 2013), but results are mixed for men. Since men tend to exhibit shorter life spans compared to women, it might be predicted that men would present higher levels of oxidative stress compared to women, perhaps in response to higher metabolic rates. But this does not seem to be the case. In two large epidemiological studies in the USA and Japan, men were found to have lower levels of peroxidative stress as measured in venous blood and urine, respectively (Sakano et al. 2009; Block et al. 2002). Greater adiposity in women was initially believed to underlie this result; however, corrections for body mass index (BMI) did not alter the difference. Body fat was not measured directly.

A study of dose-dependent responses to testosterone administration after treatment with a GnRH agonist resulted in different levels of oxidative stress in older and younger men. Younger men showed a significant decrease in oxidative stress, while older men did not (Roberts et al. 2014). However, results from these studies merit caution since many potentially relevant factors were not considered. Investigations that attempt to test the existence of a trade-off between mating effort (higher testosterone) and oxidative stress often fail to account for important factors such as energetic status, key resource restriction, and appropriate assessments of reproductive effort (Metcalfe and Monaghan 2013).

Studies using measures of whole-body oxidative stress may not detect localized or cumulative tissue-specific effects implicated in aging-related disease and dysfunction. For example, testosterone promotes greater levels of DNA damage in prostate cancer cell lines in culture (Ide et al. 2012). More tissue-specific effects of

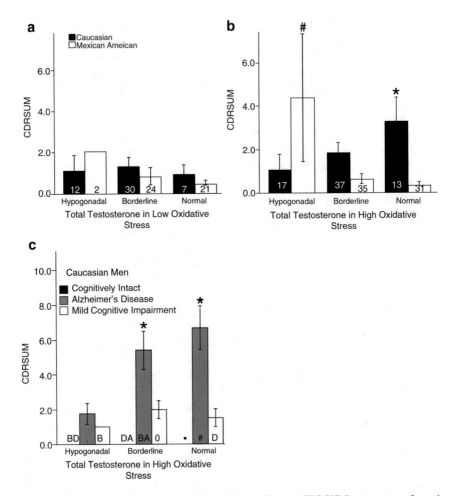

Fig. 6.5 Clinical Dementia Rating Scale Sum of Boxes Scores (CDRSUM), a measure of cognitive dysfunction, and the influence of oxidative stress and endogenous testosterone. In the high oxidative stress condition, testosterone significantly increased CDRSUM scores in Caucasian men, but significantly decreased CDRSUM scores in Mexican American men (Cunningham et al. 2014)

ROS, such as on aging-related changes in cognitive function, have also received attention. Cunningham and colleagues (Cunningham et al. 2014), for example, examined oxidative stress, testosterone levels, and level of cognitive dysfunction in normal, mildly cognitive impaired, and Alzheimer's disease patients. Their result showed that white male Alzheimer's disease patients with high testosterone and high oxidative stress (as measured by looking at levels of superoxide dismutase 1 and glutathione S-transferase alpha in venous blood) exhibited significantly higher levels of cognitive dysfunction (Fig. 6.5).

The role of oxidative stress and other forms of aging-related cellular damage is a ripe area of research for human evolutionary biologists and others interested in pleiotropic effects that may underlie men's health challenges. New technologies and the deployment of these research methods to a broader range of human populations, cultures, and ecological settings are likely to yield promising and informative results.

6.6 Men's Health and Reproductive Effort: New Directions

We argue that male-biased sex disparities in health are often grounded in life history trade-offs between maintenance and reproductive effort, with hormones acting as a pivotal agent underlying the onset of illness and disease (Bribiescas and Ellison 2007). Moreover, the broad range of human male reproductive strategies compared to those in the other great apes has the potential to allow for a greater degree of phenotypic and behavioral plasticity that can support well-being and somatic health in modern humans. This argument suggests that clinicians and public health professionals need to rethink the notion of health and illness. For example, in women, having more children has been shown to lead to significantly shorter postmenopausal life spans in some studies (Jasienska et al. 2006; Helle et al. 2002). But does this mean that childbearing should be considered an illness or health hazard? While pregnancy and lactation are not considered to be illnesses, there are life history ramifications that can have a direct bearing on women's somatic health and well-being (Jasienska 2013). Similarly, in men, hormone fluctuations in association with reproductive state (i.e., sexual maturity and fatherhood) have the potential of contributing to variation in men's health.

Biological anthropologists have employed between-population comparisons as a research strategy with great success. However, other opportunities are emerging that may augment our understanding of the interaction between male reproductive states, neuroendocrinology, phenotypic and behavioral plasticity, and morbidity and mortality. One such opportunity arises from recent increases in use of physician-administered testosterone supplementation. While hypogonadism, aging, fertility issues, and other factors may confound our understanding of the effects of testosterone supplementation on male reproductive strategies, the vast majority of men engaging in testosterone supplementation do not exhibit any symptoms of androgen deficiency (Araujo et al. 2007). It should be noted that some men receive prescribed exogenous androgen treatment by medical practitioners, others are not. However both practices are growing. The ethics and efficacy of this prescription practice are debatable and merit its own conversation. However, the downstream effects of testosterone supplementation on these men may provide a lens into the interaction of hormone-driven reproductive effort and health costs.

It can be hypothesized that testosterone supplementation may increase the probability of (1) mate-seeking behavior, (2) increased risky behavior, (3) decreased investment in offspring, or (4) decreased investment in pair-bonding. Concurrently, since testosterone is an anabolic steroid, testosterone supplementation also holds the potential to increase metabolic costs, potentially diverting resources away from immunocompetence, and perhaps overtaxing organ function that has already been degraded by aging (Muehlenbein and Bribiescas 2005). For example, men over the age of 65 were more likely to suffer cardiac infarctions within 90 days of initiating testosterone supplementation compared to control men (Finkle et al. 2014). It should be remembered, however, that most men taking testosterone supplements are from populations that already exhibit the highest and broadest ranges of testosterone levels, so conclusions should be tempered with caution (Ellison et al. 2002; Harman et al. 2001).

Investigators must also begin to move beyond measurements of mean plasma hormone levels when possible. Common sources of variation such as diurnal fluctuations are well known. However, hormones also often exhibit significant within-individual variation due to hypothalamic pulses of GnRH, CRF, receptor levels, and other regulatory agents. Researchers have hypothesized that steroid hormone production involves nonrandom pulsatile patterns that may generate an analogue-like signal that contributes to target receptor and tissue function (Thompson and Kaiser in press).

6.7   Summary

The concepts of "health" and "illness" have different meanings when viewed through the lens of evolutionary biology. For men, health is a currency that fluctuates significantly in response to energetic, social, cultural, and genetic constraints, as well as age. Moreover, it is a currency that is defined and valued differently between medical professionals and human evolutionary biologists. In strict evolutionary terms, health is only relevant to the extent that it allows animals to acquire the mates and resources needed to optimize lifetime reproductive success. During the course of human evolution, fitness trade-offs affecting men's health over the life course have likely coevolved with changes in extrinsic mortality and concurrent selection for a wide range of reproductive strategies. Seen in these evolutionary terms, many men's health issues, like prostate cancer, can be understood as a result of physiological compromises between the primacy of reproductive effort, the need for somatic maintenance, and plastic responses to ecological conditions.

To promote a fuller understanding of men's health in evolutionary medical context, human evolutionary biologists need to expand their repertoire of experimental regimens, and medical professionals need to embrace the utility of thinking about medical issues in evolutionary context. The incorporation of evolutionary and life history theory—including the concept of endocrine pleiotropy—into our understanding of health and disease is at the core of the field of evolutionary medicine (Nesse et al. 2010; Stearns et al. 2010).

6.8   Study Questions

1. What are the primary reproductive trade-offs that drive health challenges in men?

2. How do hormones affect the trade-offs between human male reproductive strategies and common health challenges?
3. What are the possible interactions between fatherhood and men's health?
4. What are the primary differences between mortality rates of younger and older men?

References

Alonso-Alvarez C, Bertrand S, Faivre B, Chastel O, Sorci G (2007) Testosterone and oxidative stress: the oxidation handicap hypothesis. Proc R Soc B Biol Sci 274(1611):819–825. doi:10.1098/rspb.2006.3764

Amundadottir LT, Sulem P, Gudmundsson J, Helgason A, Baker A, Agnarsson BA, Sigurdsson A, Benediktsdottir KR, Cazier JB, Sainz J, Jakobsdottir M, Kostic J, Magnusdottir DN, Ghosh S, Agnarsson K, Birgisdottir B, Le Roux L, Olafsdottir A, Blondal T, Andresdottir M, Gretarsdottir OS, Bergthorsson JT, Gudbjartsson D, Gylfason A, Thorleifsson G, Manolescu A, Kristjansson K, Geirsson G, Isaksson H, Douglas J, Johansson JE, Balter K, Wiklund F, Montie JE, Yu X, Suarez BK, Ober C, Cooney KA, Gronberg H, Catalona WJ, Einarsson GV, Barkardottir RB, Gulcher JR, Kong A, Thorsteinsdottir U, Stefansson K (2006) A common variant associated with prostate cancer in European and African populations. Nat Genet 38(6):652–658. doi:10.1038/ng1808

Andersson MB (1994) Sexual selection. Monographs in behavior and ecology. Princeton University Press, Princeton, NJ

Araujo AB, Esche GR, Kupelian V, O'Donnell AB, Travison TG, Williams RE, Clark RV, McKinlay JB (2007) Prevalence of symptomatic androgen deficiency in men. J Clin Endocrinol Metab 92(11):4241–4247

Austad SN (2006) Why women live longer than men: sex differences in longevity. Gend Med 3(2):79–92

Austad SN (2011) Sex differences in longevity and aging. In: Masoro EJ, Austad SN (eds) Handbook of the biology of aging, 7th edn. Academic, New York, pp 479–495

Bateman AJ (1948) Intra-sexual selection in Drosophila. Heredity 2:349–368

Beckerman S, Valentine P (2002) Cultures of multiple fathers: the theory and practice of partible paternity in lowland South America. University Press of Florida, Gainesville

Bell DL, Breland DJ, Ott MA (2013) Adolescent and young adult male health: a review. Pediatrics 132(3):535–546. doi:10.1542/peds.2012-3414

Block G, Dietrich M, Norkus EP, Morrow JD, Hudes M, Caan B, Packer L (2002) Factors associated with oxidative stress in human populations. Am J Epidemiol 156(3):274–285. doi:10.1093/aje/kwf029

Bonduriansky R, Maklakov A, Zajitschek F, Brooks R (2008) Sexual selection, sexual conflict and the evolution of ageing and life span. Funct Ecol 22(3):443–453. doi:10.1111/j.1365-2435.2008.01417.x

Bribiescas RG (1996) Testosterone levels among Aché hunter/gatherer men: a functional interpretation of population variation among adult males. Hum Nat 7(2):163–188

Bribiescas RG (2001) A model of male energetic trade offs is illustrated

Bribiescas RG (2005) Age-related differences in serum gonadotropin (FSH and LH), salivary testosterone, and 17-beta estradiol levels among Ache Amerindian males of Paraguay. Am J Phys Anthropol 127:114–121

Bribiescas RG (2006a) On the evolution of human male reproductive senescence: proximate mechanisms and life history strategies. Evol Anthropol 15(4):132–141

Bribiescas RG (2006b) Men: evolutionary and life history. Harvard University Press, Cambridge, MA

Bribiescas RG (2010) An evolutionary and life history perspective on human male reproductive senescence. Ann N Y Acad Sci 1204:54–64. doi:10.1111/j.1749-6632.2010.05524.x, NYAS5524 [pii]

Bribiescas RG, Ellison PT (2007) How hormones mediate trade-offs in human health and disease. In: Stearns SC, Koella JC (eds) Evolution in health and disease. Oxford University Press, Oxford, pp 77–93

Bribiescas RG, Ellison PT, Gray PB (2012) Male life history, reproductive effort, and the evolution of the genus Homo: new directions and perspectives. Curr Anthropol 53(S6):S424–S435. doi :10.1086/667538

Bronikowski AM, Altmann J, Brockman DK, Cords M, Fedigan LM, Pusey A, Stoinski T, Morris WF, Strier KB, Alberts SC (2011) Aging in the natural world: comparative data reveal similar mortality patterns across primates. Science 331 (6022):1325–1328. doi:10.1126/science.1201571 331/6022/1325 [pii]

Brown GR, Laland KN, Mulder MB (2009) Bateman's principles and human sex roles. Trends Ecol Evol 24(6):297–304. doi:10.1016/j.tree.2009.02.005

Burnham TC, Chapman JF, Gray PB, McIntyre MH, Lipson SF, Ellison PT (2003) Men in committed, romantic relationships have lower testosterone. Horm Behav 44(2):119–122

Busnot D (1712) Histoire du regne de Moulay Ismail. Edition Mercure de France

Byrnes JP, Miller DC, Schafer WD (1999) Gender differences in risk taking: a meta-analysis. Psychol Bull 125(3):367–383. doi:10.1037/0033-2909.125.3.367

Calistro Alvarado L (2010) Population differences in the testosterone levels of young men are associated with prostate cancer disparities in older men. Am J Hum Biol 22(4):449–455. doi:10.1002/ ajhb.21016

Carnes BA, Holden LR, Olshansky SJ, Witten MT, Siegel JS (2006) Mortality partitions and their relevance to research on senescence. Biogerontology 7(4):183–198. doi:10.1007/s10522-006-9020-3

Clay MM (1989) Quadruplets and higher multiple births. Clinics in developmental medicine, vol 107. Mac Keith and J.B. Lippincott, Oxford and Philadelphia

Cohn LD, Macfarlane S, Yanez C, Imai WK (1995) Risk-perception: differences between adolescents and adults. Health Psychol 14(3):217–222. doi:10.1037/0278-6133.14.3.217

Cunningham RL, Singh M, O'Bryant SE, Hall JR, Barber RC (2014) Oxidative stress, testosterone, and cognition among Caucasian and Mexican-American men with and without Alzheimer's Disease. J Alzheimers Dis. doi:10.3233/JAD-131994

Daly M, Wilson M (1983a) Sex, evolution, and behavior, 2nd edn. Wadsworth, Belmont, CA

Daly M, Wilson M (1983b) Sex, evolution, and behavior, 2nd edn. PWS, Boston, MA

Eisenberg ML, Park Y, Hollenbeck AR, Lipshultz LI, Schatzkin A, Pletcher MJ (2011) Fatherhood and the risk of cardiovascular mortality in the NIH-AARP Diet and Health Study. Hum Reprod 26(12):3479–3485. doi:10.1093/humrep/der305

Ellis BJ, Del Giudice M, Dishion TJ, Figueredo AJ, Gray P, Griskevicius V, Hawley PH, Jacobs WJ, James J, Volk AA, Wilson DS (2012) The evolutionary basis of risky adolescent behavior: implications for science, policy, and practice. Dev Psychol 48(3):598–623. doi:10.1037/a0026220

Ellison PT, Jasienska G (2009) Adaptation, health, and the temporal domain of human reproductive physiology. In: Panter-Brick C, Fuentes A (eds) Health, risk, and adversity: a contextual view from anthropology. Berghahn, New York, pp 108–128

Ellison PT, Bribiescas RG, Bentley GR, Campbell BC, Lipson SF, Panter-Brick C, Hill K (2002) Population variation in age-related decline in male salivary testosterone. Hum Reprod 17(12) :3251–3253

Emery Thompson M, Muller MN, Wrangham RW (2012) Technical note: variation in muscle mass in wild chimpanzees: application of a modified urinary creatinine method. Am J Phys Anthropol 149(4):622–627. doi:10.1002/ajpa.22157

Finch CE, Rose MR (1995) Hormones and the physiological architecture of life history evolution. Q Rev Biol 70(1):1–52

Finkel T, Holbrook NJ (2000) Oxidants, oxidative stress and the biology of ageing. Nature 408(6809):239–247

Finkle WD, Greenland S, Ridgeway GK, Adams JL, Frasco MA, Cook MB, Fraumeni JF Jr, Hoover RN (2014) Increased risk of non-fatal myocardial infarction following testosterone therapy prescription in men. PLoS One 9(1), e85805. doi:10.1371/journal.pone.0085805

Flinn MV, England BG (2003) Childhood stress: endocrinoe and immune responses to psychosocial events. In: Wilce JM (ed) Social and cultural lives of immune systems. Routledge, New York, pp 105–145

Garfield CF, Clark-Kauffman E, Davis MM (2006) Fatherhood as a component of men's health. JAMA 296(19):2365–2368. doi:10.1001/jama.296.19.2365

Gettler LT (2010) Direct male care and hominin evolution: why male-child interaction Is more than a nice social idea. Am Anthropol 112(1):7–21

Gettler LT, McDade TW, Feranil AB, Kuzawa CW (2011) Longitudinal evidence that fatherhood decreases testosterone in human males. Proc Natl Acad Sci. doi:10.1073/pnas.1105403108

Gettler LT, McDade TW, Agustin SS, Feranil AB, Kuzawa CW (2013) Do testosterone declines during the transition to marriage and fatherhood relate to men's sexual behavior? Evidence from the Philippines. Horm Behav 64(5):755–763. doi:10.1016/j.yhbeh.2013.08.019

Gray PB (2003) Marriage, parenting, and testosterone variation among Kenyan Swahili men. Am J Phys Anthropol 122(3):279–286

Gray PB, Anderson KG (2010) Fatherhood: evolution and human paternal behavior. Harvard University Press, Cambridge, MA

Gray PB, Kahlenberg SM, Barrett ES, Lipson SF, Ellison PT (2002) Marriage and fatherhood are associated with lower testosterone in males. Evol Hum Behav 23:193–201

Gray PB, Yang CF, Pope HG Jr (2006) Fathers have lower salivary testosterone levels than unmarried men and married non-fathers in Beijing, China. Proc Biol Sci 273(1584):333–339

Gray PB, Parkin JC, Samms-Vaughan ME (2007) Hormonal correlates of human paternal interactions: a hospital-based investigation in urban Jamaica. Horm Behav 52(4):499–507. doi:10.1016/j.yhbeh.2007.07.005, S0018-506X(07)00167-5 [pii]

Harman SM, Metter EJ, Tobin JD, Pearson J, Blackman MR (2001) Longitudinal effects of aging on serum total and free testosterone levels in healthy men. Baltimore Longitudinal Study of Aging. J Clin Endocrinol Metabol 86(2):724–731

Helle S, Lummaa V, Jokela J (2002) Sons reduced maternal longevity in preindustrial humans. Science 296(5570):1085

Hewlett BS (ed) (1992) Father-child relations: cultural and biosocial contexts. Foundations of Human Behavior; Aldine De Gruyter, Hawthorne, NY

Higley JD, Mehlman PT, Higley SB, Fernald B, Vickers J, Lindell SG, Taub DM, Suomi SJ, Linnoila M (1996) Excessive mortality in young free-ranging male nonhuman primates with low cerebrospinal fluid 5-hydroxyindoleacetic acid concentrations. Arch Gen Psychiatry 53(6):537–543

Hill K, Hurtado AM (1996) Ache life history: the ecology and demography of a foraging people. Foundations of Human Behavior; Aldine de Gruyter, New York

Hill K, Boesch C, Goodall J, Pusey A, Williams J, Wrangham RW (2001) Mortality rates among wild chimpanzees. J Hum Evol 40:437–450

Hrdy SB (2000) The optimal number of fathers. Evolution, demography, and history in the shaping of female mate preferences. Ann N Y Acad Sci 907:75–96

Ide H, Lu Y, Yu J, China T, Kumamoto T, Koseki T, Yamaguchi R, Muto S, Horie S (2012) Testosterone promotes DNA damage response under oxidative stress in prostate cancer cell lines. Prostate 72(13):1407–1411. doi:10.1002/pros.22492

Jasienska G (2009) Low birth weight of contemporary African Americans: an intergenerational effect of slavery? Am J Hum Biol 21(1):16–24. doi:10.1002/ajhb.20824

Jasienska G (2013) The fragile wisdom: an evolutionary view on women's biology and health. Harvard University Press, Cambridge, MA

Jasienska G, Thune I (2001) Lifestyle, hormones, and risk of breast cancer. BMJ 322(7286):586–587

Jasienska G, Nenko I, Jasienski M (2006) Daughters increase longevity of fathers, but daughters and sons equally reduce longevity of mothers. Am J Hum Biol 18(3):422–425

Jorgensen KT, Pedersen BV, Johansen C, Frisch M (2008) Fatherhood status and prostate cancer risk. Cancer 112(4):919–923. doi:10.1002/cncr.23230

Kaplan H, Hill K (1985) Hunting ability and reproductive success among male Ache foragers: preliminary results. Curr Anthropol 26(1):131–133

Kirkwood TB, Austad SN (2000) Why do we age? Nature 408(6809):233–238

Kokko H, Jennions MD (2008) Parental investment, sexual selection and sex ratios. J Evol Biol 21(4):919–948. doi:10.1111/j.1420-9101.2008.01540.x

Kruger DJ (2011) Evolutionary theory in public health and the public health of evolutionary theory. Futures 43(8):762–770. doi:10.1016/J.Futures.2011.05.019

Kruger DJ, Nesse RM (2006) An evolutionary life-history framework for understanding sex differences in human mortality rates. Hum Nat 17(1):74–97

Kuzawa CW, Gettler LT, Muller MN, McDade TW, Feranil AB (2009) Fatherhood, pairbonding and testosterone in the Philippines. Horm Behav 56(4):429–435. doi:10.1016/j.yhbeh.2009.07.010, S0018-506X(09)00166-4 [pii]

Lakshman KM, Kaplan B, Travison TG, Basaria S, Knapp PE, Singh AB, LaValley MP, Mazer NA, Bhasin S (2010) The effects of injected testosterone dose and age on the conversion of testosterone to estradiol and dihydrotestosterone in young and older men. J Clin Endocrinol Metab 95(8):3955–3964. doi:10.1210/jc.2010-0102

Levine RS, Rust G, Aliyu M, Pisu M, Zoorob R, Goldzweig I, Juarez P, Husaini B, Hennekens CH (2013) United States counties with low black male mortality rates. Am J Med 126(1):76–80. doi:10.1016/j.amjmed.2012.06.019

Medawar PB (1952) An unsolved problem of biology. Published for the College by H. K. Lewis, London

Metcalfe NB, Monaghan P (2013) Does reproduction cause oxidative stress? An open question. Trends Ecol Evol 28(6):347–350. doi:10.1016/j.tree.2013.01.015

Muehlenbein MP, Bribiescas RG (2005) Testosterone-mediated immune functions and male life histories. Am J Hum Biol 17(5):527–558

Muller M, Wrangham RW (2004) Dominance, aggression, and testosterone in wild chimpanzees: a test of the 'challenge hypothesis'. Anim Behav 67:113–123

Nesse RM, Bergstrom CT, Ellison PT, Flier JS, Gluckman P, Govindaraju DR, Niethammer D, Omenn GS, Perlman RL, Schwartz MD, Thomas MG, Stearns SC, Valle D (2010) Evolution in health and medicine Sackler colloquium: making evolutionary biology a basic science for medicine. Proc Natl Acad Sci U S A 107(Suppl 1):1800–1807. doi:10.1073/pnas.0906224106

Oberzaucher E, Grammer K (2014) The case of Moulay Ismael—fact or fancy? PLoS One 9(2), e85292. doi:10.1371/journal.pone.0085292

Partridge L, Barton NH (1993) Optimality, mutation and the evolution of ageing. Nature 362(6418):305–311. doi:10.1038/362305a0

Pereira ME, Fairbanks LA (2002) Juvenile primates: life history, development, and behavior. University of Chicago Press, Chicago

Quill TE, Lipkin M, Lamb GS (1984) Health-care seeking by men in their spouse's pregnancy. Psychosom Med 46(3):277–283

Rabin R (2006) Health disparities persist for men, and doctors ask why. The New York Times, 14 Nov 2006

Roberts CK, Chen BH, Pruthi S, Lee ML (2014) Effects of varying doses of testosterone on atherogenic markers in healthy younger and older men. Am J Physiol Regul Integr Comp Physiol 306(2):R118–R123. doi:10.1152/ajpregu.00372.2013

Sakano N, Wang DH, Takahashi N, Wang B, Sauriasari R, Kanbara S, Sato Y, Takigawa T, Takaki J, Ogino K (2009) Oxidative stress biomarkers and lifestyles in Japanese healthy people. J Clin Biochem Nutr 44(2):185–195. doi:10.3164/jcbn.08-252

Schroder J, Kahlke V, Staubach KH, Zabel P, Stuber F (1998) Gender differences in human sepsis. Arch Surg 133(11):1200–1205

Smith SR, Chhetri MK, Johanson J, Radfar N, Migeon CJ (1975) The pituitary-gonadal axis in men with protein-calorie malnutrition. J Clin Endocrinol Metab 41(1):60–69

Spratt DI, Crowley W Jr (1988) Pituitary and gonadal responsiveness is enhanced during GnRH-induced puberty. Am J Physiol 254(5 Pt 1):E652–E657

Stearns SC, Nesse RM, Govindaraju DR, Ellison PT (2010) Evolution in health and medicine Sackler colloquium: evolutionary perspectives on health and medicine. Proc Natl Acad Sci U S A 107(Suppl 1):1691–1695. doi:10.1073/pnas.0914475107

Thompson, IR. Kaiser UB (2014). GnRH pulse frequency-dependent differential regulation of LH and FSH gene expression Mol Cell Endocrinol 385(1-2): 28-35

Tobler M, Sandell MI (2009) Sex-specific effects of prenatal testosterone on nestling plasma antioxidant capacity in the zebra finch. J Exp Biol 212(Pt 1):89–94. doi:10.1242/jeb.020826

Tobler M, Sandell MI, Chiriac S, Hasselquist D (2013) Effects of prenatal testosterone exposure on antioxidant status and bill color in adult zebra finches. Physiol Biochem Zool 86(3):333–345. doi:10.1086/670194

Trivers RL (1972) Parental investment and sexual selection. In: Campbell BG (ed) Sexual selection and the descent of man 1871–1971. Aldine, Chicago, pp 136–179

Tuljapurkar SD, Puleston CO, Gurven MD (2007) Why men matter: mating patterns drive evolution of human lifespan. PLoS One 2(8), e785. doi:10.1371/journal.pone.0000785

Uchida A, Bribiescas RG, Ellison PT, Kanamori M, Ando J, Hirose N, Ono Y (2006) Age related variation of salivary testosterone values in healthy Japanese males. Aging Male 9(4):207–213

van Anders SM, Goldey KL, Conley TD, Snipes DJ, Patel DA (2012) Safer sex as the bolder choice: testosterone is positively correlated with safer sex behaviorally relevant attitudes in young men. J Sex Med 9(3):727–734. doi:10.1111/J.1743-6109.2011.02544.X

Vinogradov A (1998) Male reproductive strategy and decreased longevity. Acta Biotheor 46(2):157–160. doi:10.1023/A:1001181921303

Wang, X. T., Kruger, D. J., Wilke, 2009. "Life history variables and risk-taking propensity." Evolution and Human Behavior 30(2): 77–84

Walker R, Hill K (2003) Modeling growth and senescence in physical performance among the ache of eastern Paraguay. Am J Hum Biol 15(2):196–208

Walker R, Hill K, Kaplan H, McMillan G (2002) Age-dependency in hunting ability among the Ache of eastern Paraguay. J Hum Evol 42(6):639–657

Walker RS, Flinn MV, Hill KR (2010) Evolutionary history of partible paternity in lowland South America. Proc Natl Acad Sci 107(45):19195–19200. doi:10.1073/pnas.1002598107, 1002598107 [pii]

Wells JC (2000) Natural selection and sex differences in morbidity and mortality in early life. J Theor Biol 202(1):65–76. doi:10.1006/jtbi.1999.1044

Williams GC (1957) Pleiotropy, natural selection, and the evolution of senescence. Evolution 11:398–411

Wiren SM, Drevin LI, Carlsson SV, Akre O, Holmberg EC, Robinson DE, Garmo HG, Stattin PE (2013) Fatherhood status and risk of prostate cancer: nationwide, population-based case–control study. Int J Cancer 133(4):937–943. doi:10.1002/ijc.28057

Chapter 7
Immunity, Hormones, and Life History Trade-Offs

Michael P. Muehlenbein, Sean P. Prall, and Hidemi Nagao Peck

Abstract Immunity is an integral part of organismal life histories because it is crucial for maximizing evolutionary fitness, and because it is energetically expensive to develop, maintain, and activate. This chapter orients the reader to the roles of immunity in human life history trade-offs, and specifically the utility of sex hormones in mediating variation in immunity. Hormones are central mechanisms that contribute to the onset and timing of key life history events, fine-tune the optimal allocation of time and energy between competing functions, and in general modulate phenotypic development and variation. Here we describe the roles of testosterone, dehydroepiandrosterone, and estradiol in moderating immunocompetence from a life history perspective, illustrating how correlated changes in immunity and gonadal function reflect the manifold interactions between these two systems. The immunomodulatory actions of these hormones are complex and varied, and we attempt to provide explanations for this variation from the literature. Although our evidence comes from clinical medicine, our basic prediction is derived from life history theory: altering the hormonal milieu may result in differential susceptibility to both infectious and chronic diseases. Furthermore, the immunological costs associated with hormone supplementation are worthy of greater consideration by both clinical practitioners and evolutionary ecologists alike.

7.1 Trade-Offs and Hormones

Life history strategies are complex adaptations for survival and reproduction that require the coordinated evolution of somatic and reproductive developmental processes (Stearns 1992). A cornerstone of life history and evolutionary theory is the importance of phenotypic plasticity, or the ability of an organism to alter its morpho-logical, physiological, and behavioral phenotype in response to environmental change. Since environments and selection pressures can change rapidly, it is seldom adaptive for an organism to maintain a rigid set of phenotypes (Schlichting and Pigliucci 1998). Plasticity in response to stochastic ecological stressors, like the

M.P. Muehlenbein (✉) • S.P. Prall • H. Nagao Peck
Department of Anthropology, Indiana University Bloomington, Bloomington, IN, USA
e-mail: mpm1@indiana.edu

© Springer Science+Business Media New York 2017
G. Jasienska et al. (eds.), *The Arc of Life*, DOI 10.1007/978-1-4939-4038-7_7

presence of pathogens or available mates, represents a suite of complex adaptations that are manifested in the form of reaction norms produced by natural and sexual selection, and constrained by trade-offs under conditions of resource restriction (Sinervo and Svensson 1998). Reaction norms consist of the range of phenotypes that can be produced by a given genotype through short-term changes (for example, acclimatization to high altitude), as well as long-term adaptations. Phenotypic plasticity is limited through lineage-specific effects (i.e., the canalization of certain traits; phylogenetic constraints) as well as trade-offs. Assuming a limited supply of energy and time, organisms are required to allocate physiological resources between a number of competing functions, most notably reproduction, maintenance (i.e., survival), storage, work, and growth (Stearns 1989). Organisms will therefore be under selection to develop and maintain physiological systems that allow for the efficient distribution of resources between these functions. In a stochastic environment, those organisms that can most efficiently regulate the allocation of resources between competing traits will likely exhibit increased lifetime reproductive success.

Trade-offs involving reproduction are common, particularly in iteroparous (continually reproducing) organisms like humans that must balance investments between current and future reproductive events, as well as between survival and reproduction. This is to be expected given the central role of reproduction in life history evolution. Recent studies in reproductive ecology illustrate the flexibility of human reproduction in response to energetic conditions (Bribiescas 2001; Ellison 2003). Endocrinological mechanisms sensitive to environmental cues can facilitate modulation of reproductive effort relative to other investments. Both from a macro- and a microevolutionary perspective, hormones are central mechanisms that contribute to the onset and timing of key life history events, the optimal allocation of time and energy between competing functions, and the general modulation of phenotypic development and variation (Muehlenbein and Bribiescas 2005; Bribiescas and Ellison 2008; Muehlenbein and Flinn 2011). This is particularly true for steroids, ancient lipid-soluble molecules derived from cholesterol and shared by all vertebrates. Steroid hormones are involved in modulating behavior, metabolism, growth and development, reproduction, senescence, and immune functions, among others. Complex interaction and crosstalk between different steroid hormones (and other types of hormones) are therefore implicated in many aspects of human health.

It is inherently difficult to measure life history mechanisms and quantify trade-offs in humans, since we are unable to directly manipulate the system to produce genetically evolved response patterns that clearly produce phenotypic variation cued by specific environmental signals. But, as in most other organisms examined to date, the human neuroendocrine system is undoubtedly a central mediator of our phenotypic variation, including variation in life history traits (Finch and Rose 1995). For example, testosterone can facilitate male reproductive success by modifying behaviors (i.e., competition and sexual motivation) in addition to physical attributes (i.e., spermatogenesis, muscle anabolism, and fat catabolism). Musculoskeletal function can augment work capacity, intrasexual competition, intersexual coercion, and mate choice. However, high testosterone levels could also compromise survivorship by increasing energetic costs; such costs may explain the functional significance of the high variability in testosterone levels seen within men, and within and between populations (Bribiescas 2001; see Chap. 9 in this volume). This problem would become exacerbated in resource-limited environments.

The regulatory role of testosterone in allocating energetic resources and male reproduction also extends to the immune system. Maintaining high testosterone levels to bolster reproductive effort could reduce the amount of energy or nutrients available for energetically expensive immune responses. Individuals inhabiting high pathogen-risk environments may benefit from decreased testosterone levels to avoid immunosuppression and suspend energetically expensive anabolic functions (Muehlenbein 2008). Environmental conditions, including infection, during development may ultimately play an important role in altering baseline testosterone (and other hormone) levels as well as amount of variation experienced in adulthood. *The hypothesis that the benefits of testosterone trade off with immune function is based on the assumptions that immune functions are energetically costly, and that hormones play important roles in the regulation of immunity.* The immunomodulatory actions of these hormones are complex and varied, and altering the hormonal milieu may result in differential susceptibility to both infectious and chronic diseases.

7.2   Trade-Offs and Immunity

The immune system (see Box 7.1 and Fig. 7.1) is an excellent example of a reaction norm with both short- and long-term phenotypic plasticity in response to ecological stressors such as pathogens, allergens, and injury. Immunocompetence, or the ability to mount an effective immune response, is obviously an integral component of organismal life histories because it is crucial for maximizing evolutionary fitness. And because immunocompetence is an integral part of organismal life histories, it is predicted to be involved in physiological trade-offs with other functions (Sheldon and Verhulst 1996; Lockmiller and Deerenberg 2000; Norris and Evans 2000; Barnard and Behnke 2001). Selection for trade-offs is expected to be particularly strong under conditions of resource restriction, when development, maintenance, and activation of immune responses generate a substantial energetic burden (Sheldon and Verhulst 1996; Lockmiller and Deerenberg 2000; Schmid-Hempel 2003; Muehlenbein and Bribiescas 2005) (see Box 7.2).

Optimized immune functions should trade off with a variety of critical life history functions in humans, including growth and reproduction. In children, chronic immune activation in various conditions is associated with growth faltering, the failure to achieve normal growth potential (intestinal infections: Checkley et al. 1998; Campbell et al. 2003; Hadju et al. 1995; HIV infection: Arpadi 2000; inflammatory bowel disease: Ballinger et al. 2003). Likewise, nutrient deficiencies can have significant, long-term negative effects on the human immune system (Lunn 1991; Gershwin et al. 2000). Elevated C-reactive protein levels (a general measure of inflammation) are reported to be associated with reduced gains in height across 3 months in Tsimane children of Amazonian Bolivia (McDade et al. 2008). Boys in Nepal with high levels of acute-phase proteins (other proteins also associated with inflammatory states) have demonstrated stunted growth (Panter-Brick et al. 2000). Similar associations between childhood immune activation and decreased growth have been found in British children (Panter-Brick et al. 2004).

Clearly the literature points to associations between growth reduction and increased immune activation, consistent with expectations of life history theory. Illness during development may also delay menarche, as was the case for a sample of Danish women infected with *Helicobacter pylori* (Rosenstock et al. 2000) and in Guatemalan women with intestinal infections (Khan et al. 1996). Earlier menopause might also result from chronic immune activation (Cramer et al. 1983; Dorman et al. 2001). *Trade-offs between immunity and reproduction can also be identified by observing correlated changes in hormonal mechanisms responsible for the manifold interactions between these two systems.* This is particularly the case for testosterone, estradiol, and dehydroepiandrosterone.

Box 7.1: Major Mechanisms of Human Immunity

Although a comprehensive review of the human immune system is beyond the scope of this discussion (see Paul 2008), here we offer a minute summary to orient the reader (Fig. 7.1). This is meant *only* to illustrate the complexity of the immune system's dynamic responses. Typically, the human immune system is organized into two primary components innate (constitutive, nonspecific) and adaptive (acquired, specific). Innate responses include rapid mechanisms to block and eliminate foreign particles from host invasion, such as anatomical barriers, basic health behaviors, resident bacteria, humoral factors (e.g., lysozyme), and cells like neutrophils, monocytes, macrophages, basophils, mast cells, eosinophils, and natural killer cells. These cells exhibit a number of functions, from phagocytosis and cytokine secretion to chemotaxis and antigen processing and presentation. Lactoferrin, transferrin, heat shock proteins, and other soluble factors possess a variety of antimicrobial functions. The complement system includes enzymes that function to eliminate microorganisms by promoting inflammatory responses, lysis of foreign cells, and mediation of phagocytosis.

Secondary immune responses during subsequent exposures are facilitated through acquired immune mechanisms that typically involve lymphocytes (both T and B cells). B cells, produced from stem cells in bone marrow, represent antibody-mediated (humoral) immunity that involves the secretion of antibodies or "immunoglobulins" (i.e., IgG, IgM, IgA, IgD, and IgE). Antibodies neutralize pathogens and their products, block binding of parasites to host cells, induce complement activation, promote cellular migration to sites of infection, and enhance phagocytosis, among other actions. T cells, which develop in the thymus, represent cellular immunity. Different subsets of T cells are identified by their surface markers (CD numbers) that regulate cellular activation and adhesion. Cytotoxic T cells (CD8) destroy infected host cells via perforin and lysis. Suppressor T cells downregulate T cell

(continued)

Box 7.1: (continued)

responses after infection. Helper T cells (CD4) secrete cytokines and activate B cells to secrete antibodies. Cytokines are glycoproteins that perform a variety of functions such as regulation of cell growth and development. Single cytokines can have multiple functions, multiple cytokines can have similar functions, some cytokines work together to facilitate single functions, and some cytokines have opposite functions to one another.

CD4 helper T cells are generally differentiated into major subsets depending on the type of cytokine produced. For example, Th-1 cytokines include, among others, interferon gamma (IFNγ), tumor necrosis factor alpha (TNFα), and various interleukins (IL-1β, IL-2, IL-3, IL-12, etc.). These cytokines activate macrophages, neutrophils, and natural killer cells, mediate inflammatory responses and cellular immunity (T cells), promote cytotoxicity toward tumor cells, and enhance chemotaxis of leukocytes. Th-2, anti-inflammatory cytokines include many interleukins (IL-4, IL-5, IL-6, etc.) that induce humoral immunity and antibody production (B cells), eosinophil activation, mast cell degranulation, goblet cell hyperplasia, mucin production, and intestinal mastocytosis (resulting in histamine release). Despite the fact that Th-1 and Th-2 cytokines act antagonistically to one another, both are usually present within the host at any given time, although during infection one phenotype may predominate. Other Th cell types include Th-17, Tregs, Th-3 and possibly others. Clearly, single measures of immunity are not capable of capturing the complexity of such a response.

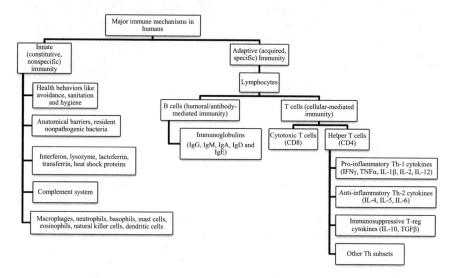

Fig. 7.1 Major immune mechanisms in humans. An illustrated summary of the complexity of the immune system's dynamic responses. For an explanation of the basic components (see Box 1). For a more comprehensive review of the human immune system (see Paul 2008). Modified from Muehlenbein (2010)

Box 7.2: Human immunity Is Energetically Expensive

In humans, prolonged energy and nutrient restriction as well as intense physical exercise can lead to immunosuppression (Chandra and Newberne 1977; Gershwin et al. 1984; Chandra 1992; Kumae et al. 1994; Pedersen and Toft 2000; Field et al. 2002); conversely, supplementation with calories, micro- and macronutrients can offset age-related declines in immunity (Wouters-Wesseling et al. 2005). The physical and psychological stress of physical exertion associated with elite athletic competitions or military training has been shown to be associated with increased incidence of upper respiratory tract infections (Peters and Bateman 1983; Nieman et al. 1990; Gomez-Merino et al. 2005). Acute infection in adult humans can cause abnormal protein loss—greater than 1 g per kilogram of body weight per day (Scrimshaw 1992). In humans, the rapid, constant turnover of T and B cells is very likely to be energetically demanding (Macallan et al. 2004, 2005).

Severe perturbations like sepsis, burns, trauma, and surgery are associated with a 25–55 % increase in resting metabolic rate compared with that in healthy subjects, as well as a reduction in body weight and total body protein (Arturson 1978; Long 1977; Kreymann et al. 1993; Frankenfield et al. 1994; Biolo et al. 1997; Carlson et al. 1997; Uehara et al. 1999; Genton and Pichard 2011), and an increase in nitrogen excretion (Carlson et al. 1997; Hasselgren and Fischer 1998). Fever typically results in a 7–15 % increase in resting metabolic rate for every 1 °C rise in body temperature (Barr et al. 1922; Roe and Kinney 1965; Elia 1992). Even in the absence of fever, resting metabolic rate is elevated during infection. For example, in a sample of 25 nonfebrile young men naturally infected with respiratory tract pathogens, resting metabolic rate was elevated by 14 % compared to samples taken after convalescence (Muehlenbein et al. 2010). Further research is needed to investigate changes in metabolic rates of adult humans during illnesses of varying severities and with different states of energy balance.

7.3   Testosterone and Immunity

Testosterone's immunomodulatory actions have usually been described as suppressive, although the results of a multitude of studies using a variety of host species are surprisingly mixed (see Muehlenbein and Bribiescas 2005 for review). In vitro experiments suggest that testosterone can increase suppressor T cell populations (Weinstein and Bercovich 1981), reduce resistance against oxidative damage (Alonso-Alvarez et al. 2007), reduce T-helper cell function (Grossman et al. 1991; Wunderlich et al. 2002), inhibit cytokine (Daynes and Araneo 1991; Grossman 1995) and antibody production (Olsen and Kovacs 1996), and impair natural killer cell and macrophage activity (Straub and Cutolo 2001). Testosterone may alter the

CD4+/CD8+ T-cell ratio in favor of CD8+ cells (Olsen et al. 1991; Weinstein and Bercovich 1981), and also favor the development of Th1 cytokines (Daynes et al. 1991; Giltay et al. 2000). *There is no reason, however, to believe a priori that testosterone should affect all aspects of immunity equally.*

Results of in vivo studies of the relationship of testosterone levels to immune status in humans are equivocal. A majority of studies conducted on healthy participants reveal few associations between testosterone and immunity. For example, in a large sample of healthy military men, Granger et al. (2000) found no association between serum testosterone levels and T or B lymphocytes, although testosterone and IgA levels were negatively correlated. No association between testosterone and IgA was identified in a smaller study of young adults (van Anders 2010). In a study of healthy male athletes, there were no associations found between testosterone and the cytokines IL-6 or IL-1β (FitzGerald et al. 2012). In a sample of 94 healthy young adults with very detailed exclusion criteria and a multi-sample collection regime, salivary testosterone levels were actually directly (positively) related to a functional measure of innate immunity (the capability of lysozyme, antibodies, complement and cells in saliva to lyse and inhibit growth of pathogenic bacteria; see Muehlenbein et al. 2011; Prall et al. 2011). Of course, variation in sampling regime, assays or laboratory conditions may explain some of the differences between studies. It is also critical to limit conclusions based on single measures of immunity, as this obviously may not accurately reflect functional immunity in terms of the ability to fight pathogens as a coordinated system (Sheldon and Verhulst 1996; Westneat and Birkhead 1998; Norris and Evans 2000). *Assays must be utilized that represent functional, integrated, biologically relevant measures of different immune pathways* (Boughton et al. 2011; Demas et al. 2011).

It is also likely that host condition and energy availability play central roles in the immunomodulatory actions of testosterone. In healthy individuals with high resource availability and relatively low energy expenditure, the immunological costs of maintaining high testosterone levels could be negated. During a disease state, in contrast, when immune functions are upregulated, those with higher testosterone (or those whom are less efficient at lowering their testosterone level; see below) may pay higher additional energetic costs and thus exhibit higher morbidity. For example, in a population of Honduran men naturally infected with *Plasmodium vivax*, those with higher testosterone during infection had significantly higher levels of malaria parasitemia (Muehlenbein et al. 2005). These men also had elevated cortisol levels during peak illness compared to recovery or to age-matched healthy controls. It seems very likely that the hormonal milieu, of which testosterone is only a small part, and including the stress endocrine axis, affects the course and outcome of infection. Glucocorticoids may play a larger role in immunoregulation than does testosterone (Turnbull and Rivier 1999).

The assumption that testosterone is globally immunosuppressive—a common, but unsupported idea in the literature—is obviously inappropriate. *Under certain conditions, testosterone's actions on immunity may in fact be beneficial.* Testosterone may actually help to prevent certain forms of immunopathology (Burger and Dayer 2002). For example, testosterone suppresses circulating immune complexes during malarial infection, which may help prevent immunopathological effects of this disease (Coleman et al. 1982). Testosterone may prevent the production of excess cytokines that might otherwise lead to tissue damage during meningitis and rheumatoid arthritis (Beutler and Cerami 1988; Waage et al. 1989). Testosterone might also function to redistribute immune cells to different parts of the body during infection (Braude et al. 1999).

Another line of evidence that testosterone is involved in mediating trade-offs between reproduction and immunity lies in its demonstrated responsiveness to illness, injury, and immune activation. Testosterone levels typically decrease in response to illness, and often correspond to the severity of perturbation (Spratt et al. 1993). Muehlenbein et al. (2005) identified lowered testosterone levels in Honduran men naturally infected with *Plasmodium vivax* compared with age-matched healthy controls. Similarly, in a sample of 25 nonfebrile young men naturally infected with respiratory tract pathogens, testosterone levels were lowered by an average of 30 % compared to those measured after recovery (Muehlenbein et al. 2010).

Variation in testosterone, and possibly other hormones, during illness may act as a physiological mechanism to decrease energy investment in reproductive effort (Muehlenbein and Bribiescas 2005; Muehlenbein 2008). This would be expected to be particularly important in high disease-risk environments and during times of limited energetic resources. Not only would depressed testosterone levels during immune activation work to limit energetic investment in energetically expensive anabolic functions, but it would also function to prevent immunosuppression by the higher testosterone levels that would be present otherwise (Folstad and Karter 1992; Wedekind and Folstad 1994; Sheldon and Verhulst 1996; Muehlenbein 2008). *Measuring changes in other hormone levels, including estrogens and leptin, during illness and throughout convalescence would be informative.*

7.4 Dehydroepiandrosterone and Immunity

As with testosterone, there has been a substantial amount of research on the immunological effects of dehydroepiandrosterone (DHEA). DHEA is a regnantoid hormone produced in the zona reticularis of the adrenal glands. DHEA and its sulfated ester DHEAS are implicated in a number of important physiological and behavioral functions. They appear to inhibit several innate immune processes, including inflammatory (Young et al. 1999; Coutinho et al. 2007) and complement responses (McLachlan et al. 1996). While this might help to ameliorate some chronic disorders, it could also increase the likelihood of impaired defense against infections. However, this liability appears to be counterbalanced by a stimulatory effect on adaptive immunity, including the development of lymphocytes (Daynes et al. 1990), particularly helper T cell activity (Suzuki et al. 1991), and proliferation of peripheral blood mononuclear cells (Sakakura et al. 2006). It is possible that DHEA also facilitates the production of Th2 over Th1 cytokines (Powell and Sonnenfeld 2006). DHEA has also been implicated in increasing Treg cytokine production (Auci et al. 2007; Coles et al. 2005).

DHEA may enhance immune responses against influenza (Corsini et al. 2006), malaria (Kurtis et al. 2001), leishmaniasis (Galindo-Sevilla et al. 2007), intestinal helminthes (Coutinho et al. 2007), and HIV (Wisniewski et al. 1993). Given the diversity of immune responses responsible for controlling such infections, however, *it is likely inappropriate to generalize DHEA's immunostimulatory effects.* Its effects may depend, in part, on the relative concentration of other hormones present. For example, in a population of 25 young men with nonfebrile acute respiratory tract infection, the ratio of DHEA to testosterone was higher during illness than after complete recovery (Prall and Muehlenbein 2011). We argue that elevated DHEA relative to testosterone might facilitate immune processes, and that a reversal of the DHEA/testosterone ratio following convalescence would downregulate immunity to prevent autoimmune reactions and bias energy expenditure towards other functions, like reproduction. These endocrine responses presumably are adaptive shifts to modulate allocations toward more immediate needs.

7.5 Estrogen and Immunity

Estradiol and other estrogens appear to be immunostimulatory. Higher circulating estrogen levels in women compared to men may help explain why females typically exhibit higher CD4+ helper T cell Th-2 cytokine responses (Bijlsma et al. 1999), greater B cell function (Soucy et al. 2005), lowered rates of cellular apoptosis (Grimaldi et al. 2002), enhanced cellular proliferation (Cutolo et al. 2005), and greater antibody secretion (Straub 2007; Cutolo et al. 2012), all of which may translate into lower morbidity and mortality from infectious diseases (Whitacre 2001). 17-beta estradiol is associated with increased immunoglobulin and cytokine levels (Olsen and Kovacs 1996; Cutolo et al. 2006). Estrogens have been shown to upregulate the production of antioxidant enzymes (Vina et al. 2006) that may decrease oxidative damage to mitochondrial DNA (Borras et al. 2007) and protect against the oxygen radicals produced by inflammatory stress (Asaba et al. 2004). Moreover, estrogens exhibit immunoprotective and anti-inflammatory properties following trauma and severe blood loss (Angele et al. 2001; Knoferl et al. 2001) and they (in contrast to testosterone, which exacerbates) protect against neuronal damage during hypoxia associated with ischemic stroke in rats (Nishino et al. 1998).

Women are naturally exposed to varying levels of estrogens as a result of cyclical variation throughout the menstrual cycle, very high levels throughout pregnancy, and a relative absence following menopause. Such variation may have important life history outcomes (Abrams and Miller 2011). Elevated levels of estrogens during ovulation and pregnancy, for example, may promote implantation and maintenance of pregnancy through anti-inflammatory (Th2) effects and temporary suppression of cell-mediated immunity (Whitacre et al. 1999; Whitacre 2001) as well as innate responses (Wira et al. 2010). Elevated progesterone levels during pregnancy appear to inhibit cytokine production (Golightly et al 2011). Therefore, during times of particularly heavy investment in female reproduction, there appears to be less investment in immunity. This appears to change when estrogens fall prior to and around menopause and there is an increase in cytokine responses (Pfeilschifter et al. 2002). But in the absence of estrogens in postmenopausal women, immune functions can become significantly impaired (Giglio et al. 1994).

Elevated levels of estrogens may contribute to the higher prevalence of autoimmune diseases seen in women (Tanriverdi et al. 2003; Straub 2007; Cutolo and Straub 2009). These disorders represent a leading cause of death and serious disability in young and middle-aged women in the USA (Cantorna and Mahon 2004), and the incidence in women compared to men is increasing significantly (Chighizola and Meroni 2012). Oral contraceptive users are at higher risk of inflammatory bowel diseases (Khalili et al. 2012) and systemic lupus erythematosus (Bernier et al. 2009). *The effects of hormone replacement therapy on health measures predicted by life history trade-offs are of critical consideration today.*

7.6   Costs and Benefits of Hormone Therapy and Supplementation

Hormone supplementation is used clinically to treat a variety of conditions. One of the most well-studied examples is estrogen therapy in women, which is often used to treat menopausal symptoms. Estrogen therapy during the menopausal transition has been shown to substantially reduce the risk of osteoporosis. It is prescribed primarily for menopausal symptoms including hot flashes (or "flushes"), insomnia, and irritability; it may also improve mood, cognitive status, and memory (NAMS 2012; Wharton et al. 2011) (see Chap. 9 in this volume). However, hormone therapy (estrogen, or estrogen in combination with progesterone) in older women has been implicated in some large clinical trials with an increased risk of blood clots, stroke, and breast cancer (Stuenkel et al. 2012). The role that estrogen plays in the risk of cardiovascular disease in older women remains controversial; this hormone, like testosterone, clearly is associated with complex physiological trade-offs that are still poorly understood.

Androgenic anabolic steroids are often used to increase quality of life and strength in both men and women (Emmelot-Vonk et al. 2008; Bhasin et al. 2010). Testosterone has been used to increase libido and improve mood (Monga et al. 2002; Gray et al 2005; Knapp et al. 2008; Panay et al. 2010), although results can be mixed (Kenny et al. 2004). Testosterone has also been used to improve memory and some measures of depression (Cherrier et al. 2001; Pope et al. 2003). Intramuscular injections of testosterone enanthate following severe burn injury can ameliorate protein catabolism, amino acid efflux, and loss of lean body mass (Ferrando et al. 2001). Similar results have been found using Oxandrolone, a synthetic derivative of testosterone, in pediatric burn patients (Tuvdendorj et al. 2011), and administration to a large sample of adult burn patients resulted in a significant reduction in mortality (Pham et al. 2008). Androgenic anabolic steroids can also ameliorate cachexia associated with cancer, renal and hepatic failure, chronic obstructive pulmonary disorder, muscular dystrophy, trauma following major surgery and anemia associated with leukemia or kidney failure (Mendenhall et al. 1995; Ferreira et al 1998; Basaria et al. 2001; Orr and Fiatarone 2004).

It is estimated that 6.5 million men in the USA will develop symptomatic, clinically recognized androgen deficiency (including lowered mood, energy and libido) by 2025 (Araujo et al. 2007). Most men with androgen deficiency either do not seek treatment for it, or are asymptomatic (Hall et al. 2008). Regardless of the cutoff values used to diagnose low testosterone, the availability of treatments and advertising by drug companies have increased. *The long-term effects of testosterone supplementation on specific aspects of health, including immune function, are largely unknown.* This problem is compounded by an increasing incidence of the use of anabolic androgenic steroids and other ergogenic (performance-enhancing) drugs for athletic enhancement or improvement of appearance (Cohen et al. 2007). The problem is not limited to professional athletes; particularly, worrisome is the dramatic rise in illegal steroid use in high school students (Calfee and Fadale 2006).

Steroid abuse in otherwise healthy individuals clearly can cause significant physical and psychological damage. These effects include a variety of conditions, from altered testicular function (Torres-Calleja et al. 2001) and acne (Walker and Adams 2009) to liver failure (Ishak 1981) and heart failure (Achar et al. 2010). Psychological effects (e.g., depression, psychosis, violence, aggression, impulsiveness, etc.) can be quite severe (Pope and Katz 1994; Bahrke et al. 1996; Beaver et al. 2008). The legal (clinical) and illegal (recreational) use of anabolic steroids has also been linked to an increased risk of prostate cancer (Shaneyfelt et al. 2000; Gaylis et al. 2005), although some studies have identified no such links (Roddam et al. 2008; Drewa and Chlosta 2010). However, the responsiveness of prostate cancer to treatments using androgen receptor inhibitors, GnRH agonists and antagonists, and even surgical castration do support an association between testosterone and prostate cancer severity and progression (Denmeade and Isaacs 2002).

The effects of testosterone supplementation on human immunity are not well investigated. In the entire volume on testosterone supplementation by Nieschlag et al. (2012), immunological consequences are mentioned only sporadically and briefly, and results of studies cited have yielded mixed results. Varying doses of testosterone do not appear to affect lymphocyte counts or viral load in HIV-infected men (Bhasin et al. 2000) and women (Choi et al. 2005; Looby et al. 2009). Testosterone treatment decreased CD4+ cell count in one study of postmenopausal women (Zofkova et al. 1995). In another study of otherwise healthy young men, there were no effects of testosterone enanthate on C-reactive protein levels (Singh et al. 2002), whereas Klinefelter's (XXY) syndrome patients have been shown to exhibit decreases in antibody levels and T cell counts following treatment with testosterone, although the percentage of CD8+ cells increased (Kocar et al. 2000). Similarly mixed results were identified by Muehlenbein and Bhasin (2012): of 52 healthy men ages 60–75 years, those who received monthly intramuscular injections of 600 mg of testosterone enanthate for 5 months showed increases in monocyte and neutrophil percentages but lowered eosinophil and lymphocyte percentages. As stated before, testosterone clearly does not affect all aspects of immunity equally, even as a result of clinically controlled supplementation.

There has also been an increased usage of DHEA by the American public as a dietary supplement in recent years (Baulieu et al. 2000). DHEA may influence metabolism and body composition, particularly through its conversion to testosterone and estradiol (Villareal and Holloszy 2004). Although other studies have identified no such relationships between body condition and DHEA level (Callies et al 2001; Percheron et al. 2003), its use as an anti-obesity agent continues to grow (Ip et al. 2011). DHEA is also purported to ameliorate some measures of depression (Wolkowitz et al. 1999) and to increase libido (Arlt et al. 1999). However, given its role as a prohormone, there are likely many other risks to supplement use, including breast cancer (Gordon et al. 1990) and ovarian cancer (Helzlsouer et al. 1995); the magnitude of risk associated with this supplement remains unknown.

Like testosterone, DHEA supplementation does not appear to affect lymphocyte counts or viral load in HIV-infected individuals (Rabkin et al. 2006; Abrams et al. 2007). Some studies have shown that DHEA supplementation may increase immune response to vaccine (Araneo et al. 1995), whereas other studies have found no such effects (Danenberg et al. 1997). DHEA may increase NK cell activity and other cellular responses in elderly recipients (Khorram et al. 1997; Casson et al. 1993), although other studies have revealed no change in these measures (Kohut et al. 2003).

The effects of hormone supplementation on the immune system require much more research to determine if the benefits of hormone therapy truly outweigh the costs. A simple prediction based on life history theory is that alterations in the hormonal mechanisms responsible for facilitating trade-offs between immune and other functions will result in dysregulation of this balanced system. Future analyses must include detailed effects of androgens and estrogens in men and women, utilizing various functional measures of adaptive immunity in a variety of experimental regimes: during health and illness of varying severity, and in people experiencing varying degrees of energy flux. *Trade-offs between immunity and other functions may only become apparent under certain conditions, or during particular critical windows at certain points in the life course.*

7.7   Summary

Phenotypic plasticity in response to stochastic ecological stressors like pathogens represents a suite of complex adaptations, and our immune system epitomizes a reaction norm that allows for adaptation to pathogens, allergens, and injury. Because immune responses presumably generate a substantial energetic burden, optimization of immunity during illness should result in decreased energetic investment in other functions, including growth and reproduction. It should be possible to indirectly observe such trade-offs by measuring correlated changes in hormones, since endocrine mechanisms are sensitive to environmental cues that can otherwise facilitate modulation of immunity relative to reproductive effort and other investments.

Testosterone, DHEA, and estradiol all appear to have complex immunomodulatory actions. Whereas testosterone's actions have usually been hypothesized to be suppressive, results of studies addressing this premise are surprisingly mixed. The same can be said for the possible immunostimulatory actions of DHEA. Estradiol may also play an important role in moderating risks of both infectious and autoimmune diseases. In short, the fluctuating, complex hormonal milieu may affect the course and outcome of disease directly through actions on immune effector mechanisms, as well as indirectly through adaptive shifts in life history allocation decisions. Although hormone supplementation clearly has beneficial actions under certain conditions, its effects on human immunity are not well investigated. Long-term augmentation of these hormonal mediators of life history trade-offs may impose significant costs on immunity against both infectious and chronic diseases.

References

Abrams ET, Miller EM (2011) The roles of the immune system in women's reproduction: evolutionary constraints and life history trade-offs. Am J Phys Anthropol 146(S53):134–154

Abrams DI, Shade SB, Couey P, McCune JM, Lo J, Bacchetti P, Chang B, Epling L, Liegler T, Grant RM (2007) Dehydroepiandrosterone (DHEA) effects on HIV replication and host immunity: a randomized placebo-controlled study. AIDS Res Hum Retroviruses 23:77–85

Achar S, Rostamian A, Narayan SM (2010) Cardiac and metabolic effects of anabolic-androgenic steroid abuse on lipids, blood pressure, left ventricular dimensions, and rhythm. Am J Cardiol 106(6):893–901

Alonso-Alvarez C, Bertrand S, Faivre B, Chastel O, Sorci G (2007) Testosterone and oxidative stress: the oxidation handicap hypothesis. Proc R Soc B 274:819–825

Angele MK, Knoferl MW, Ayala A, Bland KI, Chaudry IH (2001) Testosterone and estrogen differently effect Th1 and Th2 cytokine release following trauma-haemorrhage. Cytokine 16:22–30

Araneo B, Dowell T, Woods ML, Daynes R, Judd M, Evans T (1995) DHEAS as an effective vaccine adjuvant in elderly humans. Proof-of-principle studies. Ann N Y Acad Sci 774:232–248

Araujo AB, Esche GR, Kupelian V, O'Donnell AB, Travison TG, Williams RE, Clark RV, McKinlay JB (2007) Prevalence of symptomatic androgen deficiency in men. J Clin Endocrinol Metab 92(11):4241–4247

Arlt W, Callies F, van Vlijmen JC, Koehler I, Reincke M, Bidlingmaier M, Huebler D, Oettel M, Ernst M, Schulte HM et al (1999) Dehydroepiandrosterone replacement in women with adrenal insufficiency. N Engl J Med 341(14):1013–1020

Arpadi SM (2000) Growth failure in children with HIV infection. J Acquir Immune Defic Syndr 25:S37–S42

Arturson MGS (1978) Metabolic changes following thermal injury. World J Surg 2:203–214

Asaba K, Iwasaki Y, Yoshida M, Asai M, Oiso Y, Murohara T, Hashimoto K (2004) Attenuation by reactive oxygen species of glucocorticoid suppression on proopiomelanocortin gene expression in pituitary corticotroph cells. Endocrinology 145:39–42

Auci D, Kaler L, Subramanian S, Huang Y, Frincke J, Reading C, Offner H (2007) A new orally bioavailable synthetic androstene inhibits collagen-induced arthritis in the mouse: androstene hormones as regulators of regulatory T cells. Ann N Y Acad Sci 1110:630–640

Bahrke MS, Yesalis CE, Wright JE (1996) Psychological and behavioural effects of endogenous testosterone and anabolic-androgenic steroids. An update. Sports Med 22(6):367

Ballinger AB, Savage MR, Sanderson IR (2003) Delayed puberty associated with inflammatory bowel disease. Pediatr Res 53:205–210

Barnard CJ, Behnke JM (2001) From psychoneuroimmunology to ecological immunology: life history strategies and immunity trade-offs. In: Ader R, Felton DL, Cohen N (eds) Psychoneuroimmunology. Academic, San Diego, pp 35–47

Barr DP, Russell MD, Cecil L, Du Boise EF (1922) Clinical calorimetry XXXII: temperature regulation after the intravenous injections of protease and typhoid vaccine. Arch Intern Med 29:608–634

Basaria S, Wahlstrom JT, Dobs AS (2001) Anabolic-androgenic steroid therapy in the treatment of chronic diseases. J Clin Endocrinol Metab 86:5108–5117

Baulieu EE, Thomas G, Legrain S, Lahlou N, Roger M, Debuire B, Faucounau V, Girard L, Hervy MP, Latour F et al (2000) Dehydroepiandrosterone (DHEA), DHEA sulfate, and aging: contribution of the DHEAge Study to a sociobiomedical issue. Proc Natl Acad Sci U S A 97(8):4279–4284

Beaver KM, Vaughn MG, DeLisi M, Wright JP (2008) Anabolic-androgenic steroid use and involvement in violent behavior in a nationally representative sample of young adult males in the United States. Am J Public Health 98(12):2185

Bernier M, Mikaeloff Y, Hudson M, Suissa S (2009) Combined oral contraceptive use and the risk of systemic lupus erythematosus. Arthritis Care Res 61(4):476–481

Beutler B, Cerami A (1988) Cachectin (tumor necrosis factor): a macrophage hormone governing cellular metabolism and inflammatory response. Endocr Rev 9:57–66

Bhasin S, Storer TW, Javanbakht M, Berman N, Yarasheski KE, Phillips J, Dike M, Sinha-Hikim I, Shen R, Hays RD, Beall G (2000) Testosterone replacement and resistance exercise in HIV-infected men with weight loss and low testosterone levels. JAMA 283:763–770

Bhasin S, Cunningham GR, Hayes FJ, Matsumoto AM, Snyder PJ, Swerdloff RS, Montori VM (2010) Testosterone therapy in men with androgen deficiency syndromes: an Endocrine Society clinical practice guideline. J Clin Endocrinol Metab 95(6):2536–2559

Bijlsma JW, Cutolo M, Masi AT, Chikanza IC (1999) The neuroendocrine immune basis of rheumatic diseases. Immunol Today 20:298–301

Biolo G, Toigo G, Ciocchi B, Situlin R, Iscra F, Gullo A, Guarnieri G (1997) Metabolic response to injury and sepsis: changes in protein metabolism. Nutrition 13:52S–57S

Borras C, Gambini J, Vina J (2007) Mitochondrial oxidant generation is involved in determining why females live longer than males. Front Biosci 12:1008–1013

Boughton RK, Joop G, Armitage SAO (2011) Outdoor immunology: methodological considerations for ecologists. Funct Ecol 25:81–100

Braude S, Tang-Martinez Z, Taylor GT (1999) Stress, testosterone, and the immunoredistribution hypothesis. Behav Ecol 10:354–360

Bribiescas RG (2001) Reproductive ecology and life history of the human male. Yearb Phys Anthropol 33:148–176

Bribiescas RG, Ellison PT (2008) How hormones mediate tradeoffs in human health and disease. In: Stearns SC, Koella JC (eds) Evolution in health and disease, 2nd edn. Oxford University Press, New York, pp 77–94

Burger D, Dayer JM (2002) Cytokines, acute-phase proteins, and hormones: IL-1 and TNF-alpha production in contact-mediated activation of monocytes by T lymphocytes. Ann N Y Acad Sci 966:464–473

Calfee R, Fadale P (2006) Popular ergogenic drugs and supplements in young athletes. Pediatrics 117(3):e577–e589

Callies F, Fassnacht M, van Vlijmen JC, Koehler I, Huebler D, Seibel MJ, Arlt W, Allolio B (2001) Dehydroepiandrosterone replacement in women with adrenal insufficiency: effects on body composition, serum leptin, bone turnover, and exercise capacity. J Clin Endocrinol Metab 86(5):1968–1972

Campbell DI, Elia M, Lunn PG (2003) Growth faltering in rural Gambian infants is associated with impaired small intestinal barrier function, leading to endotoxemia and systemic inflammation. J Nutr 133:1332–1338

Cantorna MT, Mahon BD (2004) Mounting evidence for vitamin D as an environmental factor affecting autoimmune disease prevalence. Exp Biol Med 229(11):1136–1142

Carlson GL, Gray P, Arnold J, Little RA, Irving MH (1997) Thermogenic, hormonal and metabolic effects of intravenous glucose infusion in human sepsis. Br J Surg 84:1454–1459

Casson PR, Andersen RN, Herrod HG, Stentz FB, Straughn AB, Abraham GE, Buster JE (1993) Oral dehydroepiandrosterone in physiologic doses modulates immune function in postmenopausal women. Am J Obstet Gynecol 169(6):1536–1539

Chandra RK (ed) (1992) Nutrition and immunology. ARTS Biomedical, St. John's

Chandra RK, Newberne PM (1977) Nutrition, immunity and infection: mechanisms of interactions. Plennum, New York

Checkley W, Epstein LD, Gilman RH, Black RE, Cabrera L, Sterling CR (1998) Effects of *Crytosporidium parvum* infection in Peruvian children: growth faltering and subsequent catch-up growth. Am J Epidemiol 148:497–506

Cherrier MM, Asthana S, Plymate S, Baker L, Matsumoto AM, Peskind E, Raskind MA et al (2001) Testosterone supplementation improves spatial and verbal memory in healthy older men. Neurology 57(1):80–88

Chighizola C, Meroni PL (2012) The role of environmental estrogens and autoimmunity. Autoimmun Rev 11:A493–A501

Choi HH, Gray PB, Storer TW, Calof OM, Woodhouse L, Singh AM, Padero C et al (2005) Effects of testosterone replacement in human immunodeficiency virus-infected women with weight loss. J Clin Endocrinol Metab 90(3):1531–1541

Cohen J, Collins R, Darkes J, Gwartney D (2007) A league of their own: Demographics, motivations and patterns of use of 1,955 male adult non-medical anabolic steroid users in the United States. J Int Soc Sports Nutr 4(1):1–14

Coleman RM, Rencricca NJ, Fawcett PT, Veale MC, LoConte MA (1982) Androgen suppression of circulating immune complexes and enhanced survival in murine malaria. Proc Soc Exp Biol Med 171:294–297

Coles AJ, Thompson S, Cox AL, Curran S, Gurnell EM, Chatterjee VK (2005) Dehydroepiandrosterone replacement in patients with Addison's disease has a bimodal effect on regulatory (CD4+CD25hi and CD4+FoxP3+) T cells. Eur J Immunol 35(12):3694–3703

Corsini E, Vismara L, Lucchi L, Viviani B, Govoni S, Galli CL, Marinovich M, Racchi M (2006) High interleukin-10 production is associated with low antibody response to influenza vaccination in the elderly. J Leukoc Biol 80(2):376–382

Coutinho HM, Leenstra T, Acosta LP, Olveda RM, McGarvey ST, Friedman JF, Kurtis JD (2007) Higher serum concentrations of DHEAS predict improved nutritional status in helminth-infected children, adolescents, and young adults in Leyte, the Philippines. J Nutr 137(2):433–439

Cramer DW, Welch WR, Cassells S, Scully RE (1983) Mumps, menarche, menopause, and ovarian cancer. Am J Obstet Gynecol 147(1):1–6

Cutolo M, Straub RH (2009) Insights into endocrine-immunological disturbances in autoimmunity and their impact on treatment. Arthritis Res Ther 11(2):218

Cutolo M, Capellino S, Montagna P, Ghiorzo P, Sulli A, Villaggio B (2005) Sex hormone modulation of cell growth and apoptosis of the human monocytic/macrophage cell line. Arthritis Res Ther 7(5):R1124

Cutolo M, Capellino S, Sulli A, Serioli B, Secchi ME, Villaggio B, Straub RH (2006) Estrogens and autoimmune diseases. Ann N Y Acad Sci 1089:538–547

Cutolo M, Sulli A, Straub RH (2012) Estrogen metabolism and autoimmunity. Autoimmun Rev 11(6–7):A460–A464

Danenberg HD, Ben-Yehuda A, Zakay-Rones Z, Gross DJ, Friedman G (1997) Dehydroepiandrosterone treatment is not beneficial to the immune response to influenza in elderly subjects. J Clin Endocrinol Metab 82(9):2911–2914

Daynes RA, Araneo BA (1991) Regulation of T-cell function by steroid hormones. In: Meltzer MA, Mantovani A (eds) Cellular and cytokine networks in tissue immunity. Wiley-Liss, New York, pp 77–82

Daynes RA, Dudley DJ, Araneo BA (1990) Regulation of murine lymphokine production in vivo. II. Dehydroepiandrosterone is a natural enhancer of interleukin 2 synthesis by helper T cells. Eur J Immunol 20(4):793–802

Daynes RA, Meikle AW, Araneo BA (1991) Locally active steroid hormones may facilitate compartmentalization of immunity by regulating the types of lymphokines produced by helper T cells. Res Immunol 142:40–45

Demas GE, Zysling DA, Beechler BR, Muehlenbein MP, French SS (2011) Beyond phytohaemag-glutinin: assessing vertebrate immune function across ecological contexts. J Anim Ecol 80:710–730

Denmeade SR, Isaacs JT (2002) A history of prostate cancer treatment. Nat Rev Cancer 2(5):389–396

Dorman JS, Steenkiste AR, Foley TP, Strotmeyer ES, Burke JP, Kuller LH, Kwoh CK (2001) Menopause in type 1 diabetic women: is it premature? Diabetes 50(8):1857–1862

Drewa T, Chlosta P (2010) Testosterone supplementation and prostate cancer, controversies still exist. Acta Pol Pharm 67(5):543–546

Elia M (1992) Energy expenditure to metabolic rate. In: McKinney JM, Tucker HN (eds) Energy metabolism: tissue determinants and cellular corollaries. Raven, New York, pp 19–49

Ellison PT (2003) Energetics and reproductive effort. Am J Hum Biol 15(3):342–351

Ellison PT, Bribiescas RG, Bentley GR, Campbell BC, Lipson SF, Panter-Brick C, Hill K (2002) Population variation in age-related decline in male salivary testosterone. Hum Reprod 17(12):3251–3253

Emmelot-Vonk MH, Verhaar HJJ, Nakhai Pour HR, Aleman A, Lock TMTW, Ruud Bosch JLH, Grobbee DE, van der Schouw YT (2008) Effect of testosterone supplementation on functional mobility, cognition, and other parameters in older men: a randomized controlled trial. JAMA 299:39–52

Feldman HA, Longcope C, Derby CA, Johannes CB, Araujo AB, Coviello AD, Bremner WJ, McKinlay JB (2002) Age trends in the level of serum testosterone and other hormones in middle-aged men: longitudinal results from the Massachusetts male aging study. J Clin Endocrinol Metab 87(2):589–598

Ferrando AA, Sheffield-Moore M, Wolf SE, Herdon DN, Wolfe RR (2001) Testosterone adminis-tration in severe burns ameliorates muscle catabolism. Crit Care Med 29:1936–1942

Ferreira IM, Verrexchi IT, Nery LE, Goldstein RS, Zamel N, Brooks D, Jardim JR (1998) The influence of 6 months of oral anabolic steroids on body mass and respiratory muscles in under-nourished COPD patients. Chest 114:19–28

Field CJ, Johnson IR, Schley PD (2002) Nutrients and their role in host resistance to infection. J Leukoc Biol 71(1):16–32

Finch CE, Rose MR (1995) Hormones and the physiological architecture of life history evolution. Q Rev Biol 70:1–52

FitzGerald LZ, Robbins WA, Kesner JS, Xun L (2012) Reproductive hormones and interleukin-6 in serious leisure male athletes. Eur J Appl Physiol 112:3765–3773

Folstad I, Karter AJ (1992) Parasites, bright males and the immunocompetence handicap. Am Nat 139:603–622

Frankenfield DC, Wiles CE, Bagley S, Siegel JH (1994) Relationships between resting and total energy expenditure in injured and septic patients. Crit Care Med 22:1796–1804

Galindo-Sevilla N, Soto N, Mancilla J, Cerbulo A, Zambrano E, Chavira R, Huerto J (2007) Low serum levels of dehydroepiandrosterone and cortisol in human diffuse cutaneous leishmaniasis by Leishmania mexicana. Am J Trop Med Hyg 76(3):566–572

Gaylis FD, Lin DW, Ignatoff JM, Amling CL, Tutrone RF, Cosgrove DJ (2005) Prostate cancer in men using testosterone supplementation. J Urol 174(2):534–538

Genton L, Pichard C (2011) Protein catabolism and requirements in severe illness. Int J Vitam Nutr Res 81(2–3):143

Gershwin ME, Beach RS, Hurley LS (1984) Nutrition and immunity. Academic, New York

Gershwin ME, German JB, Keen CL (2000) Nutrition and immunology. Humana Press, Totowa, NJ

Giglio T, Imro MA, Filaci G, Scudeletti M, Puppo F, De Cecco L, Indiveri F, Costantini S (1994) Immune cell circulating subsets are affected by gonadal function. Life Sci 54(18):1305–1312

Giltay EJ, Fonk JC, von Blomberg BM, Drexhage HA, Schalkwijk C, Gooren LJ (2000) In vivo effects of sex steroids on lymphocyte responsiveness and immunoglobulin levels in humans. J Clin Endocrinol Metab 85:1648–1657

Golightly E, Jabbour HN, Norman JE (2011) Endocrine immune interactions in human parturition. Mol Cell Endocrinol 335:52–59

Gomez-Merino D, Drogou C, Chennaoui M, Tiollier E, Mathieu J, Gueznnec CY (2005) Effects of combined stress during intense training on cellular immunity, hormones and respiratory infections. Neuroimmunomodulation 12:164–172

Gordon GB, Bush TL, Helzlsouer KJ, Miller SR, Comstock GW (1990) Relationship of serum levels of dehydroepiandrosterone and dehydroepiandrosterone sulfate to the risk of developing postmenopausal breast cancer. Cancer Res 50(13):3859–3862

Granger DA, Booth A, Johnson DR (2000) Human aggression and enumerative measures of immunity. Psychosom Med 62:583–590

Gray PB, Singh AB, Woodhouse LJ, Storer TW, Casaburi R, Dzekov J, Dzekov C, Sinha-Hikim I, Bhasin S (2005) Dose-dependent effects of testosterone on sexual function, mood, and visuospatial cognition in older men. J Clin Endocrinol Metab 90(7):3838–3846

Grimaldi CM, Cleary J, Dagtas AS, Moussai D, Diamond B (2002) Estrogen alters thresholds for B cell apoptosis and activation. J Clin Invest 109:1625–1633

Grossman CJ (1995) The role of sex steroids in immune system regulation. In: Grossman CJ (ed) Bilateral communication between the endocrine and immune systems. Springer, New York, pp 1–11

Grossman CJ, Roselle GA, Mendenhall CL (1991) Sex steroid regulation of autoimmunity. J Steroid Biochem Mol Biol 40:649–659

Hadju V, Abadi K, Stephenson LS, Noor NN, Mohammed HO, Bowman DD (1995) Intestinal helminthiasis, nutritional status, and their relationship: a cross-sectional study in urban slum school children in Indonesia. Southeast Asian J Trop Med Public Health 26:719–729

Hall SA, Esche GR, Araujo AB, Travison TG, Clark RV, Williams RE, McKinlay JB (2008) Correlates of low testosterone and symptomatic androgen deficiency in a population-based sample. J Clin Endocrinol Metab 93(10):3870–3877

Harman SM, Metter EJ, Tobin JD, Pearson J, Blackman MR (2001) Longitudinal effects of aging on serum total and free testosterone levels in healthy men. Baltimore Longitudinal Study of Aging. J Clin Endocrinol Metab 86(2):724–731

Hasselgren PO, Fischer JE (1998) Sepsis: stimulation of energy-dependent protein breakdown resulting in protein loss in skeletal muscle. World J Surg 22:203–208

Havelock JC, Auchus RJ, Rainey WE (2004) The rise in adrenal androgen biosynthesis: adrenarche. Semin Reprod Med 22(4):337–347

Helzlsouer KJ, Alberg AJ, Gordon GB, Longcope C, Bush TL, Hoffman SC, Comstock GW (1995) Serum gonadotropins and steroid hormones and the development of ovarian cancer. JAMA 274(24):1926–1930

Hicks MJ, Jones JF, Thies AC, Weigle KA, Minnich LL (1983) Age-related changes in mitogen-induced lymphocyte function from birth to old age. Am J Clin Pathol 80(2):159–163

Ip EJ, Barnett MJ, Tenerowicz MJ, Perry PJ (2011) The anabolic 500 survey: characteristics of male users versus nonusers of anabolic-androgenic steroids for strength training. Pharmacotherapy 31:757–766

Ishak KG (1981) Hepatic lesions caused by anabolic and contraceptive steroids. Semin Liver Dis 1(2):116–128

Kenny AM, Fabregas G, Song C, Biskup B, Bellantonio S (2004) Effects of testosterone on behavior, depression, and cognitive function in older men with mild cognitive loss. J Gerontol A Biol Sci Med Sci 59(1):M75–M78

Khalili H, Higuchi LM, Ananthakrishnan AN, Richter JM, Fuchs CS, Chan AT (2012) Reproductive factors and risk of ulcerative colitis and Crohn's disease: results from two large prospective Cohorts of US Women. Gastroenterology 142(5):S89–S89

Khan AD, Schroeder DG, Martorell R, Haas JD, Rivera J (1996) Early childhood determinants of age at menarche in rural Guatemala. Am J Hum Biol 8(6):717–723

Khorram O, Vu L, Yen SS (1997) Activation of immune function by dehydroepiandrosterone (DHEA) in age-advanced men. J Gerontol A Biol Sci Med Sci 52(1):M1–M7

Knapp PE, Storer TW, Herbst KL, Singh AB, Dzekov C, Dzekov J, LaValley M, Zhang A, Ulloor J, Bhasin S (2008) Effects of a supraphysiological dose of testosterone on physical function, muscle performance, mood, and fatigue in men with HIV-associated weight loss. Am J Physiol Endocrinol Metab 294(6):E1135–E1143

Knoferl MW, Jarrar D, Angele MK, Ayala A, Schwacha MG, Bland KI, Chaudry IH (2001) 17 beta-estradiol normalizes immune responses in ovariectomized females after trauma-hemorrhage. Am J Physiol Cell Physiol 281:1131–1138

Kocar IH, Yesilova Z, Ozata M, Turan M, Sengul A, Ozdemir IC (2000) The effect of testosterone replacement treatment on immunological features of patients with Klinefelter's syndrome. Clin Exp Immunol 121:448–452

Kohut ML, Thompson JR, Campbell J, Brown GA, Vukovich MD, Jackson DA, King DS (2003) Ingestion of a dietary supplement containing dehydroepiandrosterone (DHEA) and andro-stenedione has minimal effect on immune function in middle-aged men. J Am Coll Nutr 22(5):363–371

Kreymann G, Grosser S, Buggisch P, Gottschall C, Matthaei S, Greten H (1993) Oxygen consumption and resting metabolic rate in sepsis, sepsis syndrome and septic shock. Crit Care Med 21:1012–1019

Kumae T, Kurakake S, Machida K, Sugawara K (1994) Effect of training on physical exercise-induced changes in non-specific humoral immunity. Jpn J Phys Fit Sports Med 43:75–83

Kurtis JD, Mtalib R, Onyango FK, Duffy PE (2001) Human resistance to *Plasmodium falciparum* increases during puberty and is predicted by dehydroepiandrosterone sulfate levels. Infect Immun 69(1):123–128

Lansoud-Soukate J, Dupont A, De Reggi ML, Roelants GE, Capron A (1989) Hypogonadism and ecdysteroid production in *Loa loa* and *Mansonella perstans* filariasis. Acta Trop 46:249–256

Lockmiller RL, Deerenberg C (2000) Trade-offs in evolutionary immunology: just what is the cost of immunity? Oikos 88:87–98

Long CL (1977) Energy balance and carbohydrate metabolism in infection and sepsis. Am J Clin Nutr 30:1301–1310

Looby SED, Collins M, Lee H, Grinspoon S (2009) Effects of long-term testosterone administration in HIV-infected women: a randomized, placebo-controlled trial. AIDS 23(8):951

Lukas WD, Campbell BC, Ellison PT (2004) Testosterone, aging, and body composition in men from Harare, Zimbabwe. Am J Hum Biol 16(6):704–712

Lunn PG (1991) Nutrition, immunity and infection. In: Schofield R, Reher DS, Bideau A (eds) The decline of mortality in Europe. Oxford University Press, New York, pp 131–145

Macallan DC, Wallace D, Zhang Y, de Lara C, Worth AT, Ghattas H, Griffin GE, Beverley PCL, Tough DF (2004) Rapid turnover of effector-memory CD4(1) T cells in healthy humans. J Exp Med 200:255–260

Macallan DC, Wallace DL, Zhang Y, Ghattas H, Asquith B, de Lara C, Worth A, Panayiotakopoulos G, Griffin GE, Tough DF, Beverley PCL (2005) B-cell kinetics in humans: rapid turnover of peripheral blood memory cells. Blood 105:3633–3640

McDade TW, Reyes-Garcia V, Tanner S, Huanca T, Leonard WR (2008) Maintenance versus growth: investigating the costs of immune activation among children in lowland Bolivia. Am J Phys Anthropol 136:478–484

McLachlan JA, Serkin CD, Bakouche O (1996) Dehydroepiandrosterone modulation of lipopolysaccharide-stimulated monocyte cytotoxicity. J Immunol 156(1):328–335

Mendenhall CL, Moritz TE, Rosell GA, Morgan TR, Nemchausky BA, Tamburro CH, Schiff ER, McClain CJ, Marsano LS, Allen JI, Samanta A (1995) Protein energy malnutrition in severe alcoholic hepatitis: diagnosis and response to treatment. The VA Cooperative Study Group. J Parenter Enteral Nutr 19:248–265

Monga M, Kostelec M, Kamarei M (2002) Patient satisfaction with testosterone supplementation for the treatment of erectile dysfunction. Syst Biol Reprod Med 48(6):433–442

Muehlenbein MP (2008) Adaptive variation in testosterone levels in response to immune activation: empirical and theoretical perspectives. Soc Biol 53:13–23

Muehlenbein MP (2010) Evolutionary medicine, immunity and infectious diseases. In: Muehlenbein MP (ed) Human evolutionary biology. Cambridge University Press, Cambridge, pp 459–490

Muehlenbein MP, Bhasin S (2012) Testosterone supplementation is associated with altered immunity in complex ways in healthy older men. Am J Hum Biol 24:236

Muehlenbein MP, Bribiescas RG (2005) Testosterone-mediated immune functions and male life histories. Am J Hum Biol 17:527–558

Muehlenbein MP, Flinn MV (2011) Patterns and processes of human life history evolution. In: Flatt T, Heyland A (eds) Mechanisms of life history evolution. Oxford University Press, Oxford, pp 153–168

Muehlenbein MP, Algier J, Cogswell F, James M, Krogstad D (2005) The reproductive endocrine response to *Plasmodium vivax* infection in Hondurans. Am J Trop Med Hyg 73:178–187

Muehlenbein MP, Hirschtick JL, Bonner JZ, Swartz AM (2010) Towards quantifying the usage costs of human immunity: altered metabolic rates and hormone levels during acute immune activation in men. Am J Hum Biol 22:546–556

Muehlenbein MP, Prall SP, Chester E (2011) Development of a noninvasive salivary measure of functional immunity in humans. Am J Hum Biol 23(2):267–268

NAMS (2012) The 2012 hormone therapy statement of the North American Menopause Society. Menopause 19(3):257–271

Nieman DC, Johanssen LM, Lee JW, Arabatzis K (1990) Infectious episodes in runners before and after the Los Angeles marathon. J Sports Med Phys Fitness 20:316–328

Nieschlag E, Behre HM, Nieschlag S (eds) (2012) Testosterone: action, deficiency, substitution, 4th edn. Cambridge University Press, Cambridge

Nishino H, Nakajima K, Kumazaki M, Fukuda A, Muramatsu K, Deshpande SB, Inubushi T, Morikawa S, Borlongan CV, Sanberg PR (1998) Estrogen protects against while testosterone exacerbates vulnerability of the lateral striatal artery to chemical hypoxia by 3-nitropropionic acid. Neurosci Res 30:303–312

Norris K, Evans MR (2000) Ecological immunology: life history trade-offs and immune defense in birds. Behav Ecol 11:19–26

Olsen NJ, Kovacs WJ (1996) Gonadal steroids and immunity. Endocr Rev 17:369–384

Olsen NJ, Watson MB, Henderson GS, Kovacs WJ (1991) Androgen deprivation induces phenotypic and functional changes in the thymus of adult mice. Endocrinology 129:2471–2476

Orr R, Fiatarone M (2004) The anabolic androgenic steroid oxandrolone in the treatment of wasting and catabolic disorders: review of efficacy and safety. Drugs 64:725–750

Panay N, Al-Azzawi F, Bouchard C, Davis SR, Eden J, Lodhi I, Rees M et al (2010) Testosterone treatment of HSDD in naturally menopausal women: the ADORE study. Climacteric 13:121–131

Panter-Brick C, Lunn PG, Baker R, Todd A (2000) Elevated acute-phase protein in stunted Nepali children reporting low morbidity: different rural and urban profiles. Br J Nutr 85:1–8

Panter-Brick C, Lunn PG, Goto R, Wright CM (2004) Immunostimulation and growth faltering in UK infants. Am J Hum Biol 16(5):581–587

Paul WE (2008) Fundamental immunology, 6th edn. Lippincott Williams & Wilkins, New York

Pedersen BK, Toft AD (2000) Effects of exercise on lymphocytes and cytokines. Br J Sports Med 34(4):246–251

Percheron G, Hogrel JY, Denot-Ledunois S, Fayet G, Forette F, Baulieu EE, Fardeau M, Marini JF (2003) Effect of 1-year oral administration of dehydroepiandrosterone to 60- to 80-year-old individuals on muscle function and cross-sectional area: a double-blind placebo-controlled trial. Arch Intern Med 163(6):720–727

Perrini S, Natalicchio A, Laviola L, Belsanti G, Montrone C, Cignarelli A, Minielli V, Grano M, De Pergola G, Giorgino R et al (2004) Dehydroepiandrosterone stimulates glucose uptake in human and murine adipocytes by inducing GLUT1 and GLUT4 translocation to the plasma membrane. Diabetes 53(1):41–52

Peters EM, Bateman ED (1983) Ultramarathon running and upper respiratory tract infections. S Afr Med J 64:582–584

Pfeilschifter J, Köditz R, Pfohl M, Schatz H (2002) Changes in proinflammatory cytokine activity after menopause. Endocr Rev 23(1):90–119

Pham TN, Klein MB, Gibran NS, Arnoldo BD, Gamelli RL, Silver GM, Jeschke MG, Finnerty CC, Tompkins RG, Herndon DN (2008) Impact of oxandrolone treatment on acute outcomes after severe burn injury. J Burn Care Res 29:902–906

Pope HG, Katz DL (1994) Psychiatric and medical effects of anabolic-androgenic steroid use: a controlled study of 160 athletes. Arch Gen Psychiatry 51(5):375

Pope HG, Cohane GH, Kanayama G, Siegel AJ, Hudson JI (2003) Testosterone gel supplementation for men with refractory depression: a randomized, placebo-controlled trial. Am J Psychiatry 160:105–111

Powell JM, Sonnenfeld G (2006) The effects of dehydroepiandrosterone (DHEA) on in vitro spleen cell proliferation and cytokine production. J Interferon Cytokine Res 26(1):34–39

Prall S, Muehlenbein M (2011) The ratio of salivary testosterone to dehydroepiandrosterone changes throughout recover from respiratory tract infections in men: implications for understanding hormone-mediated immunity. Am J Hum Biol 23:273

Prall S, Blanchard S, Hurst D, Ireland E, Lewis C, Martinez L, Rich A, Singh E, Taboas C, Muehlenbein M (2011) Salivary measures of testosterone and functional innate immunity are directly associated in a sample of healthy young adults. Am J Phys Anthropol S52:243

Rabkin JG, McElhiney MC, Rabkin R, McGrath PJ, Ferrando SJ (2006) Placebo-controlled trial of dehydroepiandrosterone (DHEA) for treatment of nonmajor depression in patients with HIV/AIDS. Am J Psychiatry 163(1):59–66

Roddam AW, Allen NE, Appleby P, Key TJ (2008) Endogenous sex hormones and prostate cancer: a collaborative analysis of 18 prospective studies. J Natl Cancer Inst 100(3):170–183

Roe CF, Kinney JM (1965) The caloric equivalent of fever. II. Influence of major trauma. Ann Surg 161:140–147

Rosenstock SJ, Jorgensen T, Andersen LP, Bonnevie O (2000) Association of helicobacter pylori infection with lifestyle, chronic disease, body-indices, and age at menarche in Danish adults. Scand J Public Health 28(1):32–40

Sakakura Y, Nakagawa Y, Ohzeki T (2006) Differential effect of DHEA on mitogen-induced proliferation of T and B lymphocytes. J Steroid Biochem Mol Biol 99(2–3):115–120

Schlichting C, Pigliucci M (1998) Phenotypic evolution: a reaction norm perspective. Sinaur, Sunderland, MA

Schmid-Hempel P (2003) Variation in immune defense as a question of evolutionary ecology. Proc R Soc Lond B 270:357–366

Scrimshaw NS (1992) Effect of infection on nutritional status. Proc Natl Sci Counc Repub China B 16:46–64

Shahabuddin S, Al-Ayed I, Gad El-Rab MO, Qureshi MI (1998) Age-related changes in blood lymphocyte subsets of Saudi Arabian healthy children. Clin Diagn Lab Immunol 5(5):632–635

Shaneyfelt T, Husein R, Bubley G, Mantzoros CS (2000) Hormonal predictors of prostate cancer: a meta-analysis. J Clin Oncol 18(4):847

Sheldon BC, Verhulst S (1996) Ecological immunology: costly parasite defenses and trade-offs in evolutionary ecology. Trends Ecol Evol 11:317–321

Sinervo B, Svensson E (1998) Mechanistic and selective causes of life history trade-offs and plasticity. Oikos 83:432–442

Singh AB, Hsia S, Alaupovic P, Sinha-Hikim I, Woodhouse L, Buchanan TA, Shen R, Bross R, Berman N, Bhasin S (2002) The effects of varying doses of T on insulin sensitivity, plasma lipids, apolipoproteins, and C-reactive protein in healthy young men. J Clin Endocrinol Metab 87:136–143

Soucy G, Boivin G, Labrie F, Rivest S (2005) Estradiol is required for a proper immune response to bacterial and viral pathogens in the female brain. J Immunol 174:6391–6398

Spratt DI, Cox P, Orav J, Moloney J, Bigos T (1993) Reproductive axis suppression in acute illness is related to disease severity. J Clin Endocrinol Metab 76:1548–1554

Stearns S (1989) Trade-offs in life-history evolution. Funct Ecol 3:259–268

Stearns S (1992) The evolution of life histories. Oxford University Press, New York

Straub RH (2007) The complex role of estrogens in inflammation. Endocr Rev 28(5):521–574

Straub RH, Cutolo M (2001) Involvement of the hypothalamic-pituitary-adrenal/gonadal axis and the peripheral nervous system in rheumatoid arthritis: Viewpoint based on a systemic pathogenetic role. Arthritis Rheum 44:493–507

Stuenkel CA, Gass ML, Manson JE, Lobo RA, Pal L, Rebar RW, Hall JE (2012) A decade after the Women's Health Initiative—the experts do agree. J Clin Endocrinol Metab 97(8):2617–2618

Suzuki T, Suzuki N, Daynes RA, Engleman EG (1991) Dehydroepiandrosterone enhances IL2 production and cytotoxic effector function of human T cells. Clin Immunol Immunopathol 61:202–211

Tanriverdi F, Silveira LF, MacColl GS, Bouloux PM (2003) The hypothalamic-pituitary-gonadal axis: immune function and autoimmunity. J Endocrinol 176:293–304

Torres-Calleja J, Gonzalez-Unzaga M, DeCelis-Carrillo R, Calzada-Sanchez L, Pedron N (2001) Effect of androgenic anabolic steroids on sperm quality and serum hormone levels in adult male bodybuilders. Life Sci 68(15):1769–1774

Turnbull AV, Rivier CL (1999) Regulation of the hypothalamic-pituitary-adrenal axis by cytokines: actions and mechanisms of action. Physiol Rev 79:1–71

Tuvdendorj D, Chinkes DL, Zang XJ, Suman OE, Aarsland A, Ferrando A, Kulp GA, Jeschke MG, Wolfe RR, Herndon DN (2011) Long-term oxandrolone treatment increases muscle protein net deposition via improving amino acid utilization in pediatric patients 6 months after burn injury. Surgery 149:645–653

Uehara M, Plank LD, Hill GL (1999) Components of energy expenditure in patients with severe sepsis and major trauma: a basis for clinical care. Crit Care Med 27(7):1295–1302

van Anders SM (2010) Gonadal steroids and salivary IgA in healthy young women and men. Am J Hum Biol 22(3):348–352

Villareal DT, Holloszy JO (2004) Effect of DHEA on abdominal fat and insulin action in elderly women and men: a randomized controlled trial. JAMA 292(18):2243–2248

Vina J, Sastre J, Pallardo FV, Gambini J, Borras C (2006) Role of mitochondrial oxidative stress to explain the different longevity between genders: protective effect of estrogens. Free Radic Res 40:1359–1365

Waage A, Halstensen A, Shalaby R, Brandtzaeg P, Kierulf P, Espevik T (1989) Local production of tumor necrosis factor alpha, interleukin 1, and interleukin 6 in meningococcal meningitis. Relation to the inflammatory response. J Exp Med 170:1859–1867

Walker J, Adams B (2009) Cutaneous manifestations of anabolic–androgenic steroid use in athletes. Int J Dermatol 48(10):1044–1048

Wedekind C, Folstad I (1994) Adaptive or nonadaptive immunosuppression by sex-hormones. Am Nat 143:936–938

Weinstein Y, Bercovich Z (1981) Testosterone effects on bone marrow, thymus and suppressor T cells in the (NZB x NZW) F1 mice: its relevance to autoimmunity. J Immunol 126:998–1002

Westneat DF, Birkhead TR (1998) Alternative hypotheses linking the immune system and mate choice for good genes. Proc R Soc Lond B 265:1065–1073

Wharton W, Baker LD, Gleason CE, Dowling M, Barnet JH, Johnson S, Carlsson C, Craft S, Asthana S (2011) Short-term hormone therapy with transdermal estradiol improves cognition for postmenopausal women with Alzheimer's disease: Results of a randomized controlled trial. J Alzheimers Dis 26(3):495–505

Whitacre CC (2001) Sex differences in autoimmune disease. Nat Immunol 2:777–780

Whitacre CC, Reingold SC, O'Looney PA (1999) A gender gap in autoimmunity. Science 283:1277–1278

Wira CR, Fahey JV, Ghosh M, Patel MV, Hickey DK, Ochiel DO (2010) Sex hormone regulation of innate immunity in the female reproductive tract: the role of epithelial cells in balancing reproductive potential with protection against sexually transmitted pathogens. Am J Reprod Immunol 63:544–565

Wisniewski TL, Hilton CW, Morse EV, Svec F (1993) The relationship of serum DHEA-S and cortisol levels to measures of immune function in human immunodeficiency virus-related illness. Am J Med Sci 305(2):79–83

Wolkowitz OM, Reus VI, Keebler A, Nelson N, Friedland M, Brizendine L, Roberts E (1999) Double-blind treatment of major depression with dehydroepiandrosterone. Am J Psychiatry 156(4):646–649

Wouters-Wesseling W, Vos AP, van Hal M, De Groot LCPGM, van Stavern WA, Bindels JG (2005) The effect of supplementation with an enriched drink on indices of immune function in frail elderly. J Nutr Health Aging 9:281–286

Wunderlich F, Benten WP, Lieberherr M, Guo Z, Stamm O, Wrehlke C, Sekeris CE, Mossmann H (2002) Testosterone signaling in T cells and macrophages. Steroids 67:535–538

Young DG, Skibinski G, Mason JI, James K (1999) The influence of age and gender on serum dehydroepiandrosterone sulphate (DHEA-S), IL-6, IL-6 soluble receptor (IL-6 sR) and transforming growth factor beta 1 (TGF-beta1) levels in normal healthy blood donors. Clin Exp Immunol 117(3):476–481

Zofkova I, Kancheva RL, Hampl R (1995) A decreasing CD4+/CD8+ ratio after one month of treatment with stanozol in postmenapausal women. Steroids 60:430–433

Chapter 8
Thyroid Disorders at Midlife: An Evolutionary Perspective

Lynnette Leidy Sievert

Abstract This chapter examines thyroid disease from an evolutionary perspective to better understand how adaptive physiological changes can sometimes be associated with deleterious consequences. Thyroid diseases are more common in women than in men, in part because of the hormonal and immunological changes associated with pregnancy. Pregnancy is a critical time for thyroid function because thyroid hormones are essential for fetal growth and development. Maternal hormonal changes and changes in maternal immune function during pregnancy may contribute to thyroid dysfunction during the postpartum period and later life. The connection between estrogen and thyroid function is adaptive when estrogen levels rise with pregnancy; however, the same estrogen–thyroid connection may be deleterious during menopause when estrogen levels fall. This chapter also considers health consequences related to the mismatch between the environment in which thyroid biology evolved and the current, iodine-supplemented diet. When the body is stressed by low environmental iodine or by increased iodine needs associated with pregnancy, the hypothalamic–pituitary–thyroid axis demonstrates a number of effective adaptations to maintain adequate thyroid hormone levels. The many ways in which the human body can adapt suggest that low levels of iodine were characteristic of the environment through much of human evolution.

8.1 Introduction

Thyroid disease is a broad term applied to a number of disorders that include goiters caused by iodine deficiency, hyperthyroidism (over active thyroid), hypothyroidism (underactive thyroid), and postpartum thyroiditis. Thyroid diseases are more common in women than in men (Glinoer 2004; Hollowell et al. 2002; Vanderpump et al. 1995) and diagnoses often occur around the time of pregnancy (Roti and Emerson 1992) or menopause (Weetman 1997). About 5/1000 women are diagnosed with hyperthyroidism and about 3/1000 women are diagnosed with hypothyroidism

L.L. Sievert (✉)
Department of Anthropology, University of Massachusetts Amherst,
Machmer Hall, Amherst, MA 01003, USA
e-mail: leidy@anthro.umass.edu

© Springer Science+Business Media New York 2017
G. Jasienska et al. (eds.), *The Arc of Life*, DOI 10.1007/978-1-4939-4038-7_8

(Lazarus and Mestman 2013). Why are women more vulnerable to thyroid dysfunction than men? Why would pregnancy and midlife be the two life stages when thyroid diseases are of increased concern? The following review explores these questions within the framework of evolutionary medicine.

In particular, this chapter applies evolutionary thinking to examine how adaptive physiological changes can sometimes be associated with deleterious consequences (Nesse 2008; Stearns et al. 2008). Thyroid function will be examined across the lifespan to identify connections between life events and life transitions, and to distinguish between fitness trade-offs and thyroid pathologies. For example, pregnancy is a particularly critical time with respect to thyroid function because thyroid hormones are essential for fetal somatic growth and brain development (Bogin 1999). To ensure that the fetus has adequate levels of thyroid hormone during the first trimester of pregnancy, maternal thyroid function is altered. This maternal adaptation may increase the likelihood of reproductive success, but perhaps at a cost to the mother's later thyroid health. Hypothyroidism is generally a disease of middle age, often following the menopausal transition (Dorman et al. 2002; Vanderpump et al. 1995). This chapter applies an evolutionary approach to question whether there is a physiological link between estrogen levels and thyroid function that may be adaptive during pregnancy, but deleterious after menopause.

Another application of evolutionary thinking is to connect adaptive changes in maternal immune function during pregnancy with deleterious changes in immune function during the postpartum period. The suppression of maternal immune function during pregnancy protects the fetus from maternal immune attack (Haig 1999); however, this adaptive suppression may come at a cost because some women experience a "bounce back" effect of increased immune function and greater likelihood of autoimmune attack on their own thyroid during the postpartum period (Lazarus 2005).

Also characteristic of evolutionary thinking is an appreciation of how the thyroid gland uses the same physiological processes to respond to both changing energetic demands (e.g. pregnancy) and changing levels of iodine in the environment. The thyroid gland orchestrates metabolic plasticity in response to multiple sources of stress. An evolutionary perspective directs attention to the mismatch between the low-iodine environment in which thyroid biology evolved and the current human diet and considers whether iodine-supplementation might have negative health consequences (de Benoist et al. 2008).

The approach of evolutionary medicine presumes that natural selection has shaped the hypothalamus–pituitary–thyroid axis throughout human evolution. To apply natural selection as the explanatory framework, thyroid function must meet three conditions (Langdon 2005): (1) It must be under genetic control, which is true for thyroid function (Kogai and Brent 2005; Yen and Brent 2013). (2) There must be variation in thyroid function, which is true both within and between populations (Hollowell et al. 2002; Leonard et al. 1999; Leonard et al. 2005; Sowers et al. 2003). Finally, (3) thyroid function must be associated with reproductive success, which is the topic of this chapter.

An evolutionary approach also prompts new hypotheses and new ways of thinking about the connection between health in younger and older ages, and in ancient and novel environments. Because there are many different kinds of thyroid dysfunction and numerous endocrinological and immunological factors involved in the etiology of thyroid disease, not all thyroid disease can be addressed from a single evolutionary perspective. This chapter examines normal thyroid function, how thyroid function changes during pregnancy, and thyroid dysfunction across the female lifespan. In particular, the interaction between estrogen and thyroid hormones is highlighted. Another point of interest is a consideration of how the ancient environment of evolutionary adaptedness (Bowlby 1969) may have contributed to the current human vulnerability to thyroid disorders.

8.2 Thyroid Function and Dysfunction

The thyroid gland, located at the base of the neck and anterior ventral part of the thorax, is the largest organ in the human body specialized for endocrine function (Greenspan and Baxter 1994; Parangi and Phitayakorn 2011). The butterfly-shaped gland produces iodide-containing hormones with a variety of functions. Thyroid hormones influence growth and development as well as the body's metabolism, and affect almost all tissues of the body. They stimulate metabolic rate, resulting in increased oxygen consumption, heart contractility, intestinal motility, bone remodeling, and degradation of such substances as cholesterol and other hormones (Neal 2001). Thyroid hormones also stimulate energy expenditure and thermogenesis. High levels promote the breakdown of glycogen to release energy (glucose), whereas low levels favor glycogen synthesis and energy storage. Thyroid hormones also accelerate the intestinal absorption of glucose.

Not surprisingly, persons with thyroid hormone deficiency (hypothyroidism) often feel sluggish, with decreased energy. They tend to experience cold intolerance, slowed cognition and speech, muscle cramps, dry skin, and constipation. In addition, cholesterol degradation is reduced, leading to greater risk of atherosclerosis. In contrast, hyperthyroidism, when thyroid hormone levels are too high, results in an "accelerated" person. Symptoms include inability to tolerate heat, tachycardia (fast heart rate), tremors, weight loss, increased appetite, and diarrhea. Excess thyroid hormone stimulates the catabolism of not only stored fat, but also muscle and bone tissue, leading to muscle weakness and greater risk for osteoporosis.

8.2.1 Thyroid Hormone Synthesis

Inside the thyroid gland, follicular cells synthesize and store thyroid hormone (Fig. 8.1). The thyroid makes two hormones: thyroxine (T4) and triiodothyronine (T3). Thyroxine is assembled from two modified tyrosine molecules hooked

Thyroid follicles within the thyroid gland

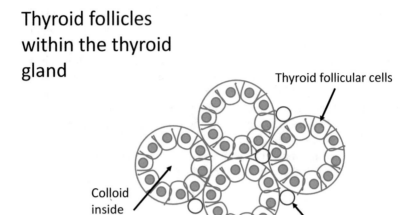

Thyroid follicular cells

Colloid inside thyroid follicle

Capillary

Fig. 8.1 Thyroid follicles inside the thyroid gland

together with four iodide atoms, hence the name T4, whereas T3 contains three iodide atoms (Kogai and Brent 2005; Prummel and Wiersinga 2002). T4 serves mainly as an available substrate for later conversion to T3, the more biologically active hormone. T4 is metabolized to T3 by deiodination (the removal of iodide) by various iodothyronine deiodinase enzymes. In the peripheral circulation, T4 is deiodinated by iodothyronine deiodinase type 1 (D1), found largely in the liver and kidney and, to a lesser extent, in the thyroid gland and muscle (Parangi and Phitayakorn 2011). T4 has a long serum half-life for a hormone: one week versus one day for T3.

Thyroid hormones travel in the bloodstream in both free and bound forms. Only the free or unbound portion is biologically active, comprising a small fraction of the total circulating hormone (only 0.4 % for T3 and 0.04 % for T4). Free hormone travels to many types of cells throughout the body, and bound hormone is transported primarily by thyroxine-binding globulin (TBG). Synthesized in the liver, TBG carries about 70 % of bound T3 and T4, the rest being transported by thyroxine-binding prealbumin (TBPA) and albumin (Greenspan and Baxter 1994; Parangi and Phitayakorn 2011).

The thyroid gland usually produces 80 % T4 and 20 % T3 (Greenspan and Baxter 1994). Some thyroid hormone is secreted immediately into the blood for use throughout the body, but much of the hormone is stored in the form of a thyroglobulin complex contained in the colloid, a gelatinous substance in the vesicles of the thyroid. Unlike most endocrine glands, the thyroid can store enough hormone to last up to several weeks. To release the stored hormone, the thyroglobulin complex passes from the colloid into the follicles where thyroid hormone is cleaved from the complex, yielding T4 and T3 molecules that flow into the bloodstream (Fig. 8.2).

The thyroglobulin complex not only provides a matrix for the synthesis of thyroid hormone, but also stores iodide. When T4 is deiodinated to form T3, the iodide

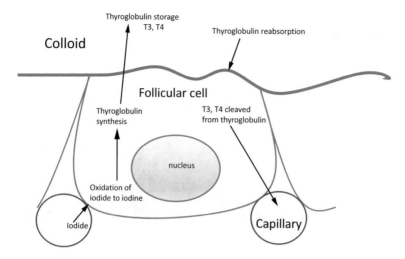

Fig. 8.2 A follicular cell lining the lumen inside a thyroid follicle

is retained in the thyroid gland for reutilization later in hormone synthesis (Greenspan and Baxter 1994). The conservation of iodide may serve as an adaptation to compensate for unpredictable access to environmental iodine. Humans need to obtain iodine (I) directly from the diet to make thyroid hormone. Iodide (I⁻) is absorbed readily into the bloodstream from the gastrointestinal tract and then removed from the blood by specialized receptors located on the surface of the thyroid cell (Kogai and Brent 2005; Parangi and Phitayakorn 2011). The concentration of free iodide in the thyroid gland can be up to 30–40 times higher than the concentration of iodide in the blood (Greenspan and Baxter 1994).

8.2.2 Thyroid Hormone Regulation

The synthesis and release of thyroid hormone is controlled and regulated by the hypothalamic–pituitary–thyroid (HPT) axis (Fig. 8.3). The hypothalamus secretes thyrotropin-releasing hormone (TRH), which stimulates the release of thyrotropin or thyroid stimulating hormone (TSH) from the anterior pituitary into the general circulation (Greenspan and Baxter 1994). As the name implies, TSH stimulates thyroid cells via TSH receptors and causes stored thyroid hormone to be released into the bloodstream (Parangi and Phitayakorn 2011). In addition, TSH stimulates all phases of iodide metabolism by increasing the uptake of iodide from the blood, the transport of iodide into thyroid follicle cells, and the biosynthesis of thyroidal peroxidase (TPO), a key enzyme used in the synthesis of thyroid hormone (Greenspan and Baxter 1994; Kogai and Brent 2005).

Thyroid hormone homeostasis is maintained through an endocrine negative feedback loop (Fig. 8.3). When blood levels of thyroid hormone become too low, the

Fig. 8.3 The
hypothalamic–pituitary–
thyroid axis feedback loop

Hypothalamic-pituitary-thyroid axis

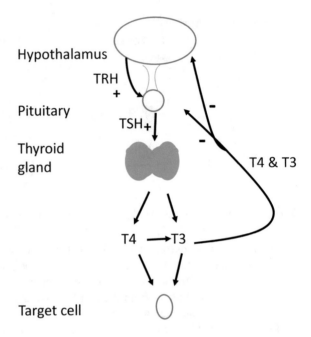

hypothalamus and pituitary respond with increased synthesis and release of TRH and TSH. The opposite occurs when blood levels become too high. High levels of thyroid hormone inhibit the release of TRH and TSH. In the hypothalamus, high levels of T3 directly inhibit TRH synthesis. T4 is converted to T3 by iodothyronine deiodinase type 2 (D2). An enzyme found mainly in neurons, D2 appears to manage the intracellular level of T3 in the central nervous system. If T4 levels rise, D2 decreases to protect the brain from excessive T3. If T4 levels fall, D2 increases to maintain intracellular levels of T3 (Greenspan and Baxter 1994).

8.2.3   Thyroid Dysfunction

Almost all thyroid disease is primary in nature, meaning due to dysfunction of the thyroid gland itself rather than the pituitary or hypothalamus. The most common cause of hypothyroidism in the USA is an autoimmune disorder called Hashimoto's disease (or Hashimoto's thyroiditis). Caused by autoantibodies that target thyroglobulin and the TPO enzyme (Dorman et al. 2002), Hashimoto's disease results in poor hormonogenesis and low circulating levels of T4 and T3 which, in turn, cause

TSH levels to rise because pituitary function is still intact. An elevated TSH level provides the most sensitive indicator of hypothyroidism. Rising TSH levels cause thyroid cells to increase in size and thyroid enlargement (goiter) can eventually develop; most patients with Hashimoto's disease have thyroid enlargement. D1 enzyme levels are also decreased in hypothyroidism to conserve T4 levels (Greenspan and Baxter 1994).

Although relatively rare in the USA, iodine deficiency is another common cause of hypothyroidism, especially in underdeveloped countries. Some drugs such as lithium can also cause hypothyroidism. Another, less common type of hypothyroidism is called atrophic thyroiditis. In this autoimmune disorder, TSH-blocking antibodies impede the usual effects of TSH on thyroid cells, causing the gland to atrophy.

The most common cause of hyperthyroidism in the USA is Graves' disease, an autoimmune disorder caused by an antibody that mimics TSH, binding directly to the thyroid gland as an "imposter" and stimulating thyroid hormone overproduction. D1 enzyme levels are increased in hyperthyroidism to accelerate T4 metabolism (Greenspan and Baxter 1994).

8.3 Thyroid Function During Pregnancy

Pregnancy is a particularly critical time with respect to thyroid function because of the rising metabolic needs of the mother who, at the same time, must provide thyroid hormone for the developing fetus, particularly during the first trimester (Glinoer 1999). When iodine intake is adequate during pregnancy, the thyroid gland increases in size (10–15 % on average) to meet the demands of increased hormone production. In this sense, pregnancy mimics hyperthyroidism (Greenspan and Baxter 1994). Although the thyroid gland is the first gland capable of hormonal activity in the developing fetus (Krakow 2008), the fetus is completely dependent on the mother's supply of thyroid hormone until about the 12th week of gestation (Greenspan and Baxter 1994). Thyroid hormones, essential for proper development of the fetus, contribute to normal body growth and proportions, as well as the formation of bone from cartilage.

Rising levels of both human chorionic gonadotropin (hCG) and estrogens, characteristic of early pregnancy, stimulate maternal thyroid hormone release. Synthesized by the placenta, with levels peaking near the end of the first trimester, hCG binds directly to TSH receptors on the mother's thyroid gland (Hershman 2004; Parangi and Phitayakorn 2011). Rising estrogen levels during early and midgestation increase TBG synthesis in the liver, thus generating more binding sites for circulating T4 and T3 and initially reducing the levels of free thyroid hormone in the mother's blood. This shift in ratio between bound and unbound thyroid hormone prompts a subsequent increase in TSH secretion from the pituitary, thereby stimulating an increase in thyroid hormone secretion (Cavalieri 1987; Green 1987). As a result of the effects of hCG and estrogens, elevated levels of free thyroid hormone are achieved and maintained for the first 20 weeks of pregnancy, with concentrations gradually returning to previous levels in the second and third trimesters (Cavalieri 1987; Lazarus 2005). Total thyroid hormone (bound and unbound) levels show an overall increase of 50–75 % from the first trimester of pregnancy until about 6 weeks postpartum (Gambert 1996).

After 12 weeks of pregnancy, the fetal hypothalamic–pituitary–thyroid system is able to function relatively independently of the mother (Bianco et al. 2002; Greenspan and Baxter 1994), at which time most maternal thyroid hormone is inactivated by the placental enzyme D3 (Glinoer 2004). Type 3 iodothyronine deiodinase (D3) is the primary iodothyronine deiodinase found in the placenta. Its activity increases with gestational age, and there is direct evidence that D3 controls transmission of maternal T4 to the fetus by converting T4 to (inactive) rT3 and by inactivating T3 so that very little free maternal hormone reaches the fetal circulation (Bianco et al. 2002; Greenspan and Baxter 1994). Notably, D3 is elevated in hyperthyroidism and decreased in hypothyroidism. The adaptive function of D3 appears to protect the developing fetus from too much maternal thyroid hormone (Bianco et al. 2002; Greenspan and Baxter 1994, p. 172).

As Bianco and colleagues (2002) explain, the relatively independent functioning of the maternal and fetal thyroid axes allows fetal thyroid hormone levels to be regulated primarily by fetal developmental needs and enables the maternal thyroid axis to respond primarily to maternal metabolic needs. Nonetheless, the fetus remains entirely dependent on the mother for its supply of iodide, which crosses the placenta easily (Krakow 2008), making adequate maternal iodine intake during pregnancy critical for fetal development.

In the presence of a healthy thyroid and an adequate supply of iodine from the diet, the mechanisms described above enable the thyroid gland to meet the increased demand for thyroid hormone for both maternal and fetal needs, especially during the first trimester, thereby maximizing the likelihood of a successful pregnancy outcome.

8.3.1   Thyroid Dysfunction Associated with Pregnancy

The ability of the maternal thyroid gland to make physiological adjustments during pregnancy can be impaired by either autoimmune thyroid disease or an inadequate supply of environmental iodine. Hypothyroidism, when the thyroid is not able to make enough hormone for both maternal and fetal needs, occurs in about 2.5 % of pregnancies (Lazarus 2005). Consequences of hypothyroidism during pregnancy include preeclampsia (characterized by high blood pressure and signs of damage to organ systems), muscle weakness, congestive heart failure, postpartum hemorrhage, miscarriage, fetal brain damage, and fetal death (Krakow 2008; Mandel 2004; Parangi and Phitayakorn 2011). In the absence of iodine deficiency, hypothyroidism during pregnancy is most commonly due to Hashimoto's disease and characterized by anti-TPO enzyme antibodies.

Pregnant women are advised to consume twice the normal amount of iodine per day to keep up with increased thyroid activity during pregnancy (Parangi and Phitayakorn 2011). When iodine intake is not adequate, the maternal thyroid gland can become stressed and can double in size (Glinoer 2004). Goiters formed during pregnancy tend to only partially regress after delivery. Urinary excretion of iodine also increases during pregnancy and can contribute to relative iodide deficiency (Lazarus 2005). Pregnancy is a particularly critical time because thyroid hormones are essential for the growth and development of the fetus. According to the World Health Organization (WHO), iodine deficiency is "the world's greatest single cause of preventable brain damage (de Benoist et al. 2008, p. 195)." The WHO identified three countries in particular (Romania, Philippines, and Nepal) where low urinary iodine levels in pregnant women were indicative of a public health problem (de Benoist et al. 2008). Cretinism, a form of dwarfism and mental retardation in off-spring, is the most devastating result of maternal iodine deficiency (Bogin 1999).

Hyperthyroidism, resulting in excessive thyroid hormone, complicates about 2 % of pregnancies (Krakow 2008; Lazarus 2005) and is commonly caused by Graves' disease (Mestman 2004). Consequences of hyperthyroidism during pregnancy include preeclampsia, preterm delivery, fetal demise, and the complete suppression of thyroid formation and function in the fetus (Krakow 2008). Sometimes women are unaware that they have Graves' disease until they experience a worsening of symptoms during pregnancy. Some symptoms of Graves' disease such as shortness of breath, palpitations, and heat intolerance are typical of pregnancy. Other symptoms not typical of pregnancy include weight loss or poor weight gain (Krakow 2008). Transient hyperthyroidism can sometimes occur as well, caused by the high levels of hCG during the first trimester of pregnancy or by hyperemesis gravidarum, a condition characterized by severe nausea and vomiting resulting in dehydration and weight loss (Speroff and Fritz 2005).

Although some physiological changes during pregnancy can increase the mother's vulnerability to thyroid disease, many pre-existing autoimmune disorders can improve during pregnancy (Barbesino and Davies 2002) because of general suppression of the mother's immune system to protect the fetus from maternal rejection (Lazarus 2005). The high levels of estrogens produced during pregnancy are also associated with a decline in the production of autoantibodies (Lazarus 2005).

8.3.2 Postpartum Thyroid Disorders

Although pre-existing autoimmune disorders can improve during pregnancy, general suppression of the maternal immune system can precipitate autoimmune changes postpartum with both short and long-term deleterious effects (Barbesino and Davies 2002). Some women develop transient hyperthyroidism or Graves' disease as an "immune rebound" response during the postpartum period (Lazarus 2005). In other words, after being suppressed during pregnancy, the immune system bounces back with postpartum hyperactivity and an increase in TSH-receptor antibodies (Gonzalez-Jimenez et al. 1993). Insomnia, nervousness, and fatigue are difficult to differentiate from normal post-pregnancy symptoms, yet approximately 2–15 % of women become hyperthyroid 6–9 months after delivery due to a new or recurrent onset of Graves' disease (Barbesino and Davies 2002; Parangi and Phitayakorn 2011). This transient condition generally lasts for several months (Parangi and Phitayakorn 2011).

Postpartum thyroid dysfunction (PPTD), also known as postpartum thyroiditis, is another type of thyroid disorder that can be precipitated by immune function changes following pregnancy. Characterized by anti-TPO antibodies and sometimes anti-thyroglobulin, PPTD has an initial hyperthyroid phase followed by a hypothyroid phase (Lazarus 2005). PPTD tends to recur with pregnancies and can lead to an increased risk of developing hypothyroidism. One follow-up study found that, at approximately 80 months after delivery, 50 % of TPO antibody-positive women with transient PPTD had developed hypothyroidism (Premawardhana et al. 2000).

8.4   Post-reproductive Thyroid Disorders

The empirical patterns for the frequency of thyroid disorders in men and women across the life course show two distinct differences between the sexes. First, women are at greater risk than men for developing thyroid disorders throughout all stages of adult life. Second, although both sexes show an increase in disease risk associated with advancing age, the magnitude of the sex difference becomes especially pronounced during the post-reproductive period.

The best-known study of the prevalence of thyroid disorders is the Whickham Survey. This survey was carried out in 1972–1974 among 2779 adults selected to closely resemble the British population. The prevalence of hypothyroidism was 1.4 % for women and <0.1 % for men (Turnbridge et al. 1977). A 20-year follow-up was carried out in 1992–1993 among 1877 survivors. The incidence rate of all cases of hypothyroidism among women, including treated cases, was 4.1 (per 1000/year). The incidence rate for men was just 0.6 (per1000/year). For women, the risk of developing hypothyroidism increased with age, with a mean age at diagnosis of 60 years (Vanderpump et al. 1995).

Hypothyroidism at midlife occurs when there is either insufficient consumption of iodine or when the immune system attacks the thyroid. In Hashimoto's disease autoantibodies target thyroglobulin and TPO (Dorman et al. 2002). In the USA, the National Health and Nutrition Examination Survey, 1988–1994 (NHANES III), indicated that 4.6 % of the US population had either clinical or mild hypothyroidism, and significantly more women than men had hypothyroidism in the 50–59 and 60–69 age groups (Hollowell et al. 2002). As shown in Fig. 8.4, women had higher median levels of TSH, which can indicate hypothyroidism. In addition, as shown in Fig. 8.5, more women had antithyroid antibodies, and the prevalence of antithyroid antibodies increased with age (Hollowell et al. 2002).

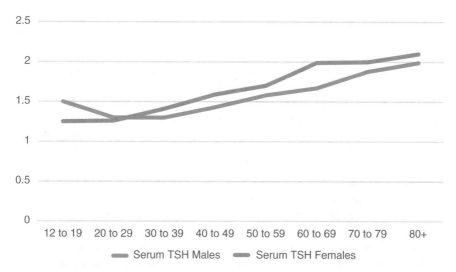

Fig. 8.4 Median serum TSH concentration by age and sex, U.S., excluding people with thyroid disease, goiter, or taking thyroid medication; NHANES III (1988–1994) (Hollowell et al. 2002)

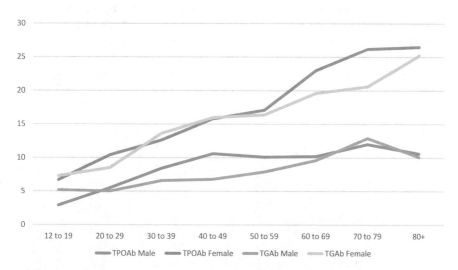

Fig. 8.5 Percentage of US population with positive antithyroid antibodies by age and sex, excluding people with thyroid disease, goiter, or taking thyroid medication; NHANES III (1988–1994) (Hollowell et al. 2002)

The higher risk for developing thyroid disorders observed in women throughout the adult life course is not surprising given both the (cumulative) risks associated with pregnancy and the greater susceptibility to autoimmune disorders evidenced for women in general (Bouman et al. 2005). The higher risk with advancing age for both sexes can be understood, in large part, in terms of the overall senescence of the immune system, which likely contributes to the greater risk of developing autoimmune disorders over time (See Chap. 7 this volume).

What would account for the precipitous rise in thyroid disorders, particularly hypothyroidism, for women during the post-reproductive period? Compared to premenopause, the postmenopausal stage of life is a relatively hypoestrogenic state, just as the postpartum period is a relatively hypoestrogenic state compared to pregnancy. The physiological pathways connecting estrogen to thyroid function, so crucial to the developing fetus during pregnancy, may lead to compromised thyroid function during the post-reproductive period when estrogen levels fall. The relationship between higher estrogen levels and thyroid stimulation during pregnancy, including increased pituitary sensitivity to TRH, synthesis of TBG by the liver, and higher circulating levels of thyroid hormone, has also been seen in women using oral contraceptives or hormone replacement therapy (NAMS 2010). The change in thyroid function associated with the post-reproductive period may result from selection for a close relationship between estrogen levels and thyroid function.

Two predicted effects of an inadequate supply of estrogen would include decreased pituitary responsiveness to TRH (via fewer TRH receptors) and decreased TBG synthesis by the liver, thereby both increasing the availability of free hormone and lowering the signal to stimulate thyroid hormonogenesis by the central axis. Additionally, the drop in estrogen levels may exacerbate women's susceptibility to autoimmune disorders. In light of the adaptive relationship between estrogen and thyroid function during pregnancy, the rise in thyroid disorders for women during the post-reproductive period may be best understood as an example of antagonistic pleiotropy (see Box 8.1). In other words, the same adaptive mechanisms that link

---

Box 8.1: Antagonistic Pleiotropy

Pleiotropy means that single genes have multiple effects. When at least one trait conferred by an allele is beneficial to the organism's survival and reproduction, and at least one trait conferred by that same allele is detrimental to the organism's survival and reproduction, those counteracting effects are known as antagonistic pleiotropy. Williams (1957) proposed that the same genes could have beneficial effects early in life, but detrimental effects later in life, resulting in senescent changes with age.

One example of antagonistic pleiotropy is testosterone activity in men. Testosterone contributes to positive effects early in life: secondary sex characteristics, the regulation of puberty, and reproductive success. However, testosterone activity may result in compromised immune function, prostate cancer, and an increased risk of cardiovascular disease later in life (Bribiescas and Ellison 2008; Finch and Rose 1995). This chapter proposes that the link between estrogen levels and thyroid function is adaptive early in the lifespan, specifically during pregnancy, but may be deleterious when estrogen levels fall during the postpartum period and after menopause.

estrogen to thyroid function during pregnancy—a life history stage under strong selective pressure—may lead to deleterious consequences later in life when selective pressures are not as great, in this case during the post-reproductive period.

8.5   Iodine Deficiency

According to the World Health Organization (WHO), two billion people have insufficient iodine intake that can result in hypothyroidism and enlargement of the thyroid gland (de Benoist et al. 2008). Insufficient iodine intake occurs throughout the world, most often in Europe and the Eastern Mediterranean as well as parts of Asia, Africa, the Andes of South America, New Guinea, and the Great Lakes region of the USA. Iodine, a mineral that originates from seawater, is essential to the functioning of the thyroid gland in all vertebrates (Kogai and Brent 2005). A lack of iodine is found in environments where the mineral has been leached from the soil mostly through erosion, deforestation, and overgrazing (Parangi and Phitayakorn 2011).

Foods rich in naturally occurring iodine include seafood, brown seaweed kelp, egg yolks, beans (red kidney, navy, pinto), leafy greens (spinach, turnip greens, Swiss chard), summer squash, rhubarb, garlic, horseradish, and sesame seeds (Parangi and Phitayakorn 2011). Many common foods, however, also contain substances known as goitrogens that can interfere with the body's ability to properly use iodine (Kogai and Brent 2005). Early field studies in highland Ecuador conducted by anthropologist Greene (1977) showed that consuming goitrogenic foods could contribute to thyroid dysfunction. Goitrogens are found in such common foods as grains (millet, maize), legumes (soybeans), tubers (turnips, cassava, sweet potatoes), leafy greens (mustard greens, kale), cruciferous vegetables (broccoli, cauliflower, cabbage, rutabaga, Brussels sprouts), onions and garlic (Lindeberg 2009; Parangi and Phitayakorn 2011). Although foods with a bitter taste, like Brussels sprouts, tend to contain goitrogens, not all goitrogens taste bitter (Trevathan 2010).

Kopp (2004) has argued that iodine deficiency and related thyroid disorders are a relatively recent phenomenon, stemming from increased iodine requirements brought about by dietary changes associated with agriculture. According to Kopp, contemporary diets, with their high starch and sugar content compared to the low carbohydrate diets of the ancestral past, raise T3 levels and subsequently raise iodine intake requirements. The increase in T3 levels from carbohydrate consumption is likely due to accelerated peripheral conversion of T4 to T3 (Danforth et al. 1979). Lindeberg (2009) has suggested that the current daily recommendations for iodine intake reflect an increased consumption of goitrogen-containing foods not often consumed or available prior to the advent of agriculture. Also, according to Lindeberg (2012), our hominin ancestors, with their lower iodine requirements, may have met those requirements by consuming the thyroid glands of animals. In contrast, Eaton and Eaton (1998) have suggested that, unlike other vitamins and minerals, iodine deficiency likely characterized the evolutionary past, noting that the availability of environmental iodine would have varied with such factors as proximity to the ocean, volcanic activity, prevailing winds, and rainfall patterns. Broadhurst et al. (2002) also argue that key nutrients, including iodine, would have been relatively less available on the African savanna and hominins living along river corridors and the sea would have had a nutritional advantage.

A consideration of the anatomy and physiology of the thyroid gland, however, suggests that humans evolved in environments of relatively low iodine availability. First, the thyroid gland has the capacity to store thyroid hormone in the form of thyroglobulin, thereby allowing secretion of thyroid hormone to continue for at least up to one month in the absence of sufficient iodine intake (Kogai and Brent 2005). Second, the thyroid gland has the capacity to retain and conserve iodide when thyroglobulin is hydrolyzed, a process that occurs at a faster rate when iodine intake is low. Third, insufficient iodine intake results in the preferential synthesis of T3, the more biologically active hormone (Greenspan and Baxter 1994); less iodide in the bloodstream results in less iodide in the thyroid gland, which shifts the ratio of hormone synthesis in favor of T3 because more iodide is needed to make T4. D1 deiodination of T4 to T3 within the thyroid gland also increases. The preferential synthesis of T3, however, can cause TSH levels to rise in response to the low circulating levels of T4. Over time, this can lead to mild or severe hypothyroidism. Fourth, the D2 enzyme, with its specialized ability to convert T4 to T3 and regulate T3 levels exclusively within neurons, appears to protect the brain from potential shortages in environmental iodine.

Taken collectively, the suite of mechanisms described above indicates an adaptive physiology designed by natural selection to compensate for environmental perturbations in the availability of iodine. Many of these same mechanisms operate to meet the increased iodine requirements associated with pregnancy. The storage of thyroid hormone, the conservation of iodide, and the preferential synthesis of T3 when iodine intake is low suggest that relatively low levels or periodically low levels of iodine may have characterized the ancestral environment throughout much of human evolution.

8.6 Iodine Excess

Today, modern diets are often supplemented with iodine in products such as iodized salt, bread, and cow's milk that help people meet the recommended dietary intake of iodine (Pearce et al. 2004; Perrine et al. 2010). Across the world there is a decline in the prevalence of iodine deficiency (de Benoist et al. 2008). However, too much of a good thing may, indeed, be too much. In 2008, the World Health Organization labeled the intake of iodine "excessive" in Armenia, Brazil, Chile, the Democratic Republic of Congo, Ecuador, Liberia, and Uganda (WHO 2008). In countries where the level of iodine intake is too high, susceptible individuals are at risk for iodine-induced thyroid dysfunction, including iodine-induced hyperthyroidism (Burgi

2010; de Benoist et al. 2008; Konno et al. 1993; Roti and Vagenakis 2013; Sang et al. 2012; Teng et al. 2011; Weetman and McGregor 1994; WHO 2008). Given that humans most likely evolved in environments characterized by low iodine availability, iodine-induced thyroid dysfunction can be viewed as an example of an evolutionary mismatch (Nesse 2008), whereby a modern health risk derives from the interaction between ancestral physiology and a novel environmental condition, such as excessive iodine intake in certain environments in which humans live today.

8.7 Conclusion

This chapter examined the physiology of the normal thyroid gland and changes in thyroid function associated with pregnancy. From an evolutionary perspective, changes that occur during pregnancy are adaptations that provide increased levels of thyroid hormone at a time when maternal needs are increased and when the fetus depends on a maternal supply of thyroid hormones. The human thyroid gland also demonstrates a suite of adaptations that enhance function in the context of low levels of environmental iodine. These adaptations support the argument that hominins evolved in ancestral environments characterized by low levels of iodine and may explain why thyroid disease is associated with excessive intake of iodine, a novel environmental concern.

Evolutionary medicine provides a way to ask, why are humans vulnerable to thyroid disease? From this perspective, thyroid disorders may result from the mismatch between ancient and novel environments. Adaptations that were selected for during hominin evolution (in what was most likely a low iodine environment) may not protect current populations from the deleterious consequences of excessive iodine intake. An evolutionary perspective also provides a way to recognize that the same trait (or suite of traits) may have beneficial effects early in the lifespan and deleterious effects later in life. The link between estrogen and thyroid function is beneficial during pregnancy, yet that same link may result in thyroid disorders after menopause.

This chapter has not been exhaustive. There are many types of thyroid disease, and the etiology is multifactorial. However, this chapter has illustrated ways in which evolutionary medicine can contribute new perspectives and new hypotheses when considering human vulnerabilities.

Acknowledgments This chapter started as a conversation with Heath Kurra many years ago. I appreciate his insight and help. I am also indebted to Diana Sherry for her encouragement and incredibly generous editing. Thanks, also, to two anonymous reviewers for added clarity.

References

Barbesino G, Davies TF (2002) The immunological mediation of Robert Graves' disease. In: Gill RG, Harmon JT, Maclaren NK (eds) Immunologically mediated endocrine diseases. Lippincott Williams & Wilkins, Philadelphia

Bianco AC, Salvatore D, Gereben B, Berry MJ, Larsen PR (2002) Biochemistry, cellular and molecular biology, and physiological roles of the iodothyronine selenodeiodinases. Endocr Rev 23(1):38–89

Bogin B (1999) Patterns of human growth, 2nd edn. Cambridge University Press, Cambridge

Bouman A, Heineman MJ, Faas MM (2005) Sex hormones and the immune response in humans. Hum Reprod Update 11(4):411–423

Bowlby J (1969) Attachment. Basic Books, New York

Bribiescas RG, Ellison PT (2008) How hormones mediate trade-offs in human health and disease. In: Stearns SC, Koella JC (eds) Evolution in health and disease, 2nd edn. Oxford University Press, Oxford

Broadhurst CL, Wang Y, Crawford MA, Cunnane SC, Parkington JE, Schmidt WF (2002) Brain-specific lipids from marine, lacustrine, or terrestrial food resources: potential impact on early African *Homo sapiens*. Comp Biochem Physiol B Biochem Mol Biol 131(4):653–673

Burgi H (2010) Iodine excess. Best Pract Res Clin Endocrinol Metab 24(1):107–115

Cavalieri RR (1987) Laboratory evaluation of thyroid status. In: Green WL (ed) The thyroid. Elsevier, New York

Danforth E, Horton ES, O'Connell M, Sims EA, Berger AG, Ingbar SH, Braverman L, Vagenakis AG (1979) Dietary-induced alterations in thyroid hormone metabolism during overnutrition. J Clin Invest 64(5):1336–1347

de Benoist B, McLean E, Andersson M, Rogers L (2008) Iodine deficiency in 2007: global progress since 2003. Food Nutr Bull 29(3):195–202

Dorman JS, Foley TP, Perry MV (2002) Epidemiology of chronic autoimmune thyroiditis and type 1 diabetes. In: Gill RG, Harmon JT, Maclaren NK (eds) Immunologically mediated endocrine diseases. Lippincott Williams & Wilkins, Philadelphia

Eaton SB, Eaton SB (1998) Evolution, diet and health. Poster presented at the 14th international congress of anthropological and ethnological sciences, Williamsburg, VA. http://www.cast. uark.edu/local/icaes/conferences/wburg/posters/sboydeaton/eaton.htm

Finch CE, Rose MR (1995) Hormones and the physiological architecture of life history evolution. Q Rev Biol 70:1–52

Gambert SR (1996) Intrinsic and extrinsic variables. age and physiologic variables. In: Braverman LE, Utiger RD (eds) Werner and Ingbar's the thyroid: a fundamental and clinical text, 7th edn. Lippincott-Raven, Philadelphia

Glinoer D (1999) What happens to the normal thyroid during pregnancy? Thyroid 9(7):631–635

Glinoer D (2004) The regulation of thyroid function during normal pregnancy: importance of the iodine nutrition status. Best Pract Res Clin Endocrinol Metab 18(2):133–152

Gonzalez-Jimenez A, Fernandez-Soto ML, Escobar-Jimenez F et al (1993) Thyroid function parameters and TSH-receptor antibodies in healthy subjects and Graves' disease patients: a sequential study before, during and after pregnancy. Thyroidology 5(1):13–20

Green WL (1987) Physiology of the thyroid gland and its hormones. In: Green WL (ed) The thyroid. Elsevier, New York

Greene LS (1977) Hyperendemic goiter, cretinism, and social organization in highland Ecuador. In: Greene LS (ed) Malnutrition, behavior, and social organization. Academic, New York

Greenspan FS, Baxter JD (1994) Basic and clinical endocrinology, 4th edn. Appleton & Lange, Norwalk, CT

Haig D (1999) Genetic conflicts in pregnancy and childhood. In: Stearns SC (ed) Evolution in health and disease. Oxford University Press, Oxford

Hershman JM (2004) Physiological and pathological aspects of the effect of human chorionic gonadotropin on the thyroid. Best Pract Res Clin Endocrinol Metab 18(2):249–265

Hollowell JG, Staehling NW, Flanders WD et al (2002) Serum TSH, T4, and thyroid antibodies in the United States population (1988 to 1994): National Health and Nutrition Examination Survey (NHANES III). J Clin Endocrinol Metab 87(2):489–499

Kogai T, Brent GA (2005) Thyroid hormones (T4, T3). In: Melmed S, Conn PM (eds) Endocrinology: basic and clinical principles, 2nd edn. Humana Press, Totowa, NJ

Konno N, Yuri K, Taguchi H et al (1993) Screening for thyroid diseases in an iodine sufficient area with sensitive thyrotrophin assays, and serum thyroid autoantibody and urinary iodide determinations. Clin Endocrinol 38(3):273–281

Kopp W (2004) Nutrition, evolution and thyroid hormone levels—a link to iodine deficiency disorders? Med Hypotheses 62(6):871–875

Krakow D (2008) Medical and surgical complications of pregnancy. In: Gibbs RS, Karlan BY, Haney AF, Nygaard IE (eds) Danforth's obstetrics and gynecology. Wolters Kluwer/Lippincott Williams & Wilkins, New York

Langdon JH (2005) The human strategy: an evolutionary perspective on human anatomy. Oxford University Press, Oxford

Lazarus JH (2005) Thyroid disorders associated with pregnancy: etiology, diagnosis, and management. Treat Endocrinol 4(1):31–41

Lazarus JH, Mestman JH (2013) Thyroid disorders during pregnancy and postpartum. In: Braverman LE, Cooper DS (eds) Werner and Ingbar's the thyroid: a fundamental and clinical text, 10th edn. Wolters Kluwer/Lippincott Williams & Wilkins, New York

Leonard WR, Galloway VA, Ivakine E et al (1999) Nutrition, thyroid function and basal metabolism of the Evenki of Central Siberia. Int J Circumpolar Health 58:281–295

Leonard WR, Snodgrass JJ, Sorensen MV (2005) Metabolic adaptation in indigenous Siberian populations. Annu Rev Anthropol 34:451–471

Lindeberg S (2009) Modern human physiology with respect to evolutionary adaptations that relate to diet in the past. In: Hublin JJ, Richards MP (eds) The evolution of hominin diets: integrating approaches to the study of palaeolithic subsistence. Springer, Berlin/Heidelberg

Lindeberg S (2012) Paleolithic diets as a model for prevention and treatment of western disease. Am J Hum Biol 24(2):110–115

Mandel SJ (2004) Hypothyroidism and chronic autoimmune thyroiditis in the pregnant state: maternal aspects. Best Pract Res Clin Endocrinol Metab 18(2):213–224

Mestman JH (2004) Hyperthyroidism in pregnancy. Best Pract Res Clin Endocrinol Metab 18(2):267–288

Neal JM (2001) How the endocrine system works. Wiley Blackwell, New Jersey

Nesse R (2008) The importance of evolution for medicine. In: Trevathan WR, Smith OE, McKenna JJ (eds) Evolutionary medicine and health: new perspectives. Oxford University Press, Oxford

North American Menopause Society (NAMS) (2010) Menopause practice: a clinician's guide, 4th edn. North American Menopause Society, Mayfield Heights, OH

Parangi S, Phitayakorn R (2011) Thyroid disease. Greenwood, Santa Barbara, CA

Pearce EN, Pino S, He X et al (2004) Sources of dietary iodine: bread, cow's milk, and infant formula in the Boston area. J Clin Endocrinol Metab 89(7):3421–3424

Perrine CG, Herrick K, Serdula MK, Sullivan KM (2010) Some subgroups of reproductive age women in the United States may be at risk for iodine deficiency. J Nutr 140(8):1489–1494

Premawardhana LDKE, Parkes AB, Ammari F et al (2000) Postpartum thyroiditis and long-term thyroid status: prognostic influence of thyroid peroxidase antibodies and ultrasound echogenicity. J Clin Endocrinol Metab 85(1):71–75

Prummel MF, Wiersinga WM (2002) Autoimmune thyroid diseases. In: Gill RG, Harmon JT, Maclaren NK (eds) Immunologically mediated endocrine diseases. Lippincott Williams & Wilkins, Philadelphia

Roti E, Emerson CH (1992) Postpartum thyroiditis. J Clin Endocrinol Metab 74(1):3–5

Roti E, Vagenakis AG (2013) Effect of excess iodide: clinical aspects. In: Braverman LE, Cooper DS (eds) Werner and Ingbar's the thyroid: a fundamental and clinical text, 10th edn. Wolters Kluwer/Lippincott Williams & Wilkins, New York

Sang Z, Wei W, Zhao N et al (2012) Thyroid dysfunction during late gestation is associated with excessive iodine intake in pregnant women. J Clin Endocrinol Metab 97(8):E1363–E1369

Sowers MF, Luborsky J, Perdue C et al (2003) Thyroid stimulating hormone (TSH) concentrations and menopausal status in women at the mid-life: SWAN. Clin Endocrinol 58:340–347

Speroff L, Fritz MA (2005) Clinical gynecologic endocrinology and infertility, 7th edn. Lippincott Williams & Wilkins, Philadelphia

Stearns SC, Nesse RM, Haig D (2008) Introducing evolutionary thinking for medicine. In: Stearns SC, Koella JC (eds) Evolution in health and disease, 2nd edn. Oxford University Press, Oxford

Teng X, Shan Z, Chen Y et al (2011) More than adequate iodine intake may increase subclinical hypothyroidism and autoimmune thryroiditis: a cross-sectional study based on two Chinese communities with different iodine intake levels. Eur J Endocrinol 164(6):943–950

Trevathan W (2010) Ancient bodies, modern lives: how evolution has shaped women's health. Oxford University Press, Oxford

Turnbridge WMG, Evered DC, Hall R et al (1977) The spectrum of thyroid disease in the community: the Whickham Survey. Clin Endocrinol 7:481–493

Vanderpump MP, Tunbridge WM, French JM et al (1995) The incidence of thyroid disorders in the community: a twenty-year follow-up of the Whickham Survey. Clin Endocrinol 43(1):55–68

Weetman AP (1997) Hypothyroidism: screening and subclinical disease. BMJ 314:1175–1178

Weetman AP, McGregor AM (1994) Autoimmune thyroid disease: further developments in our understanding. Endocr Rev 15:788–830

Williams GC (1957) Pleiotropy, natural selection, and the evolution of senescence. Evolution 11:398–411

World Health Organization (WHO) Nutrition (2008) The WHO global database on iodine deficiency. http://www.who.int/vmnis/en/en/. Accessed 25 Jan 2015

Yen PM, Brent GA (2013) Genomic and nongenomic actions of thyroid hormones. In: Braverman LE, Cooper DS (eds) Werner and Ingbar's the thyroid: a fundamental and clinical text, 10th edn. Wolters Kluwer/Lippincott Williams & Wilkins, New York

Chapter 9
Women's Health in the Post-menopausal Age

Donna J. Holmes

Abstract In recent human history, the trajectory of the human life cycle in indus-trialized cultures has undergone a profound shift. Dramatic increases (30–40 years or more) in average and maximum life expectancies have occurred, along with decreases in infant and maternal mortality rates, delays in childbearing, and fertility rate declines, plus lower mortality rates for older adults. Both the proportion and absolute numbers of older adults are growing globally, attributable in large part to the vast numbers of people born during the post-World War II "baby boom." The extension of the human life span to "extreme" old ages (80–100 years or more) has been caused by changes in culture that have outpaced biological evolution, includ-ing better nutrition, sanitation, and medical care, as well as protection against deaths from accident and exposure. An equally dramatic epidemiologic transition has accompanied this demographic change, shifting major causes of death away from infection, injury, and exposure to later-onset diseases often characterized by an extended period of disability. Women are longer-lived on average than men, and the modern "longevity bonus" means that millions of us will live a third or more of our lives in a healthy post-menopausal state. On the other hand, more older women spend their later years struggling with chronic, disabling conditions, some of which are associated with declines in estrogen and progesterone after menopause. An evo-lutionary analysis of these trends provides a valuable perspective on how the health consequences of an extended post-reproductive life span for modern women.

9.1 Introduction

Since the mid-1800s, a profound demographic shift has been underway for modern humans (Table 9.1). While this shift is occurring primarily in industrialized coun-tries, the economic and health impacts associated with it are manifesting themselves globally (Olshansky et al. 1990; Olshansky 1992; Crews 2003; Vaupel 2010). Mortality rates have slowed at all stages of the life cycle, and life expectancies (defined as the probability of living to a certain age based on statistical averages for

D.J. Holmes (✉)
Department of Biological Sciences and WWAMI Medical Education Program,
University of Idaho, MS 3051, Moscow, ID 83844-3051, USA
e-mail: djholmes@wsu.edu

© Springer Science+Business Media New York 2017
G. Jasienska et al. (eds.), *The Arc of Life*, DOI 10.1007/978-1-4939-4038-7_9

Table 9.1 Recent demographic shifts associated with the emergence of extremely long modern human life spans

The demographic shift since the Industrial Revolution. Since the mid-1800s, human populations in industrialized countries all over the world have experienced a dramatic demographic shift, including the following:
• Longer average life spans for both men and women
Example: Average life spans[a] in the USA:
1900: 45 years
2000s: 78–80+ years
Note: "Life span" denotes statistical average life expectancy (LE) from birth calculated for the US population as a whole. Some people lived longer than this average; others did not attain the average LE.
• Longer maximum life spans, and an increasing number of "centenarians" (people living to 100 years or more) worldwide
Longest-lived person on record, Jeanne Calment (a French woman), lived to 122 years (died in 1997)
Note: "Maximum life span" denotes the extreme case—the longest life span ever lived by anyone in a particular population for which reliable documentation of longevity is available.
• Longer segments of life spent in less-fertile or post-reproductive states
Average age at menopause: about 50 years
Average life span for women (LE at birth, USA): 80+ years
Years in a post-reproductive state: up to 30+ years
• Decreases in annual death rates at all ages
• Decreasing fertility rates
• Changes in most prevalent diseases and other causes of death to more chronic, aging-related diseases (epidemiologic shift)
• Changes in force of natural selection

Sources: Crews (2003), Finch and Austad (2008), Vaupel (2010) and National Center for Health Statistics (2012)

a particular population) at all ages have steadily increased (Table 9.2; Fig. 9.1). At the same time, fertility rates have dropped markedly. Increases in longevity have been accompanied by a dramatic "epidemiological transition": a shift in major causes of death away from the infectious diseases, famine, exposure, and accidents that plagued our ancestors, and toward a much higher prevalence of chronic diseases and disabilities associated with older ages. These unprecedented changes in rates of death and disease are manifestations of the impact of modern culture on human health, effects which have outpaced rates of evolutionary, and genetic change affecting the human physiological phenotype.

For modern women and men, a much longer segment of the life span that ever before is spent in a reduced-fertility or post-reproductive state. Millions of women can now expect to live 30 years or more (sometimes over a third of their lives) beyond menopause (around age 50), when menstrual cycles stop and circulating levels of estrogen and progesterone drop to a vestige of their levels during the fertile segment of life. This post-menopausal longevity bonus is undoubtedly one of the blessings of modern life. The health dividends of post-menopausal life, however,

Table 9.2 Shift in average life expectancy[a] for men and women combined in the USA since 1900

Year	Life expectancy at birth (years)	Life expectancy at age 65 (years)
1900	47.7	11.7
1935	60.9	12.7
2000	76.7	17.4

[a]Life expectancy is defined as the expected number of years of life remaining for a person of a given age, based on the statistical average life span for a particular population or country. Because it is calculated as an average, it represents the probability of living to a certain age; some people will outlive the average, while others will not attain it

Sources: Manton et al. (2006), Austad (2011), Bell and Miller (2005) and Holmes and Cohen (2014); Human Mortality Database (www.mortality.org)

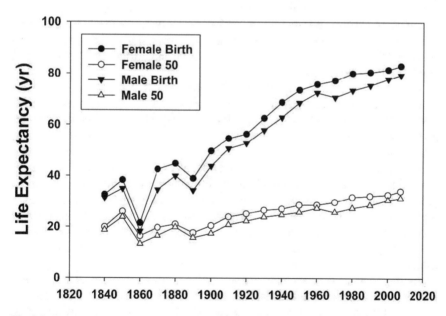

Fig. 9.1 Increases in life expectancy at birth and age 50 in females and males in Iceland. Reprinted from Austad (2011) with permission. The dip in life expectancies in Europe in the mid-1800s has been attributed to famine due to crop failure and loss of livestock, as well as epidemics of infectious diseases like smallpox, scarlet fever, and diphtheria, following the "Little Ice Age." (Appleby 1980; Guttormsson and Garðarsdóttir 2002)

come packaged with some subtle—and often unrecognized—physiological trade-offs. In many ways, postmillennial women are exploring new physiological territory as we reach old age in unprecedented numbers, and face the challenges of sustaining optimal health during the decades after menopause. The aim of this chapter is to couch these physiological trade-offs and health challenges in evolutionary terms and in the context of our ancestral physiological heritage. We also consider some of the implications of the recent shift in women's life expectancy as viewed through the lens of evolutionary medicine.

9.1.1 Biogerontology

Many of the health challenges experienced by women in the modern world at midlife and beyond can be analyzed using a conceptual framework from the evolutionary biology of aging, also referred to as "*biogerontology*" (Rose 1991) or "geroscience" (Lithgow 2013; Kennedy et al. 2014) (Table 9.3). Biogerontologists integrate methods from a variety of disciplines—evolutionary theory, demography, genetics, comparative physiology, and developmental biology—to understand the biology of human aging in the context of evolution and natural history, as well as in a broader comparative framework. The field of biogerontology incorporates *evolutionary aging theory*, a set of ideas that can be applied to understand the phenomenon of aging in terms of Darwinian evolutionary theory. Aging theory has gradually matured since the 1950s, in close association with evolutionary life-history theory (reviewed in Kirkwood and Austad 2000; Finch and Austad 2008; Cohen and Holmes 2014) (see references cited in Table 9.3).

A central idea from biogerontology is the prediction that fitness trade-offs shape organismal life histories to promote lifetime reproductive success (Kirkwood 1981; Rose 1991; Stearns 1992; Partridge and Barton 1993). Since reproduction is costly, successful populations of organisms are expected to evolve strategies that sacrifice long-term maintenance and repair of somatic (non-reproductive) structures in order to maximize lifetime reproductive effort. Differences in mortality rates are seen as the major force driving natural selection. Variation in mortality rates, in turn, is predicted to shape life-history trajectories along a continuum of short-lived, fast-reproducing vs. longer-lived, slower-reproducing organisms. As the likelihood of survival and successful reproduction declines over the course of an animal's life span, the force of natural selection wanes, and the probability of aging-related deterioration, dysfunction, and disease is expected to increase.

An important tenet of evolutionary aging theory (also referred to as "*senescence theory*") is that aging (organismal "senescence") is not expected to be adaptive. Since aging cannot directly promote the reproductive fitness of the individual, evolutionary biologists generally do not view aging as the result of a "program" directly shaped by natural selection (Williams 1966). Instead, aging is expected to arise by default, as a by-product of the *failure* of evolution to promote a longer reproductive life span and extended period of organismal integrity. Another key idea from evolu-

Table 9.3 Key concepts and processes from the biology of aging ("geroscience" or "biogerontology") and evolutionary aging theory

Aging (organismal senescence) can be defined as a deterioration of physiological function with advancing chronological age, leading to an increased probability of disease, disability, and death.
General characteristics of aging:
Aging is
• Not "programmed" or shaped directly by natural selection, hence generally not expected to serve a primary adaptive function
• Correlated in its timing to the failure of developmental or physiological events critical for reproductive success earlier in life (e.g., infertility in female mammals is linked to finite stores of oocytes formed during prenatal development)
• Often has an unpredictable trajectory for individual organisms in a given population, with different systems often deteriorating at different rates
Evolutionary processes ("ultimate mechanisms") proposed to be involved in aging:
• Increasing probability of disease, deterioration, and death with chronological age
• Waning natural selection with age, as mortality force and probability of reproduction decline
• Accumulation of mutations in older individuals in a population
• Accumulation of molecular and cellular damage with age
• Gradual failure of repair mechanisms
• Fitness trade-offs between investment in reproduction vs. maintenance and defense of somatic tissues (life-history trade-offs)
• Deleterious effects later in life of genes with beneficial effects earlier in life (antagonistic pleiotropy)
Basic biological processes ("proximate mechanisms") proposed to be involved in aging:
• Oxidative stress and other forms of damage to macromolecules, cells, and tissues (including damage to DNA)
• Mitochondrial dysregulation
• Dysregulation of transcription, cell replication, and cellular signaling patterns leading to unregulated cell growth
• Failure and dysregulation of inflammatory and autoimmune processes
• Changes in lipid metabolism
• Changes in energy metabolism
• Endocrine processes involved in growth and reproduction
• Failure of mechanisms for homeostasis and repair

References: Williams (1957), Rose (1991), Partridge and Barton (1993), Finch and Kirkwood (2000), Kirkwood and Austad (2000), Zera and Harshman (2001) and Holmes and Cohen (2014)

tionary aging theory is that genes or physiological processes that are beneficial for development or reproduction early in life may exert correlated, deleterious effects later in life. This phenomenon, "*antagonistic pleiotropy*" (Williams 1957), can help to explain why some essential endocrine and cellular processes may become dysregulated and even harmful at older ages. An additional tenet of evolutionary aging theory is that harmful mutations in a population are more likely to accumulate and be expressed later in life, as the force of natural selection wanes.

To understand the complex health issues associated with living longer lives, evolutionary biologists can compare health and life span variables for humans living in modern, industrialized cultures with those of traditional hunter-gatherer societies (see, for example, Gurven et al. 2008, 2009). Another approach is to compare clinical health variables, reproductive aging, and mortality patterns in modern people with those of nonhuman primate species living in captivity or in the wild, or those of non-primate mammalian or vertebrate species with different social lives and life spans (see, for example, Cohen 2004; Finch and Holmes 2010). Yet another comparative evolutionary approach to understanding aging and health is to study variation across species in genes, cell-signaling pathways, and hormones responsible for life span, basic aging processes, or specific aging-associated diseases in nonhuman animals, and attempting to place these functions in adaptive context in healthy, reproductive animals (Tatar et al. 2003; Ackermann and Pletcher 2008; McDonald 2014).

Evolutionary approaches like these help to identify the molecular, cellular, and physiological changes associated with variation in life span, basic aging processes, and some diseases of aging in humans (for reviews, see Kirkwood and Austad 2000; Promislow et al. 2006; Finch 2007). A continued application of evolutionary principles to specific aging-related conditions—including those experienced during and after the menopausal transition—can provide a framework for understanding specific physiological and endocrine trade-offs likely to be associated with living beyond the reproductive life span for women in the new millennium.

9.2　"Extreme Aging"

Modern humans in industrialized countries generally now live approximately 35–40 years longer on average (and up to 50 years longer at the upper extremes of life span) than even our very recent ancestors. In the USA, the average life expectancy at birth (for women and men combined) has increased from about 47 (1900) to over 76 years (2000) over the past century or so (Vaupel 2010; Austad 2011) (Tables 9.1 and 9.2; Fig. 9.1). In Japan (the longest-lived country in the world) and some other developed nations, average life spans for both sexes now exceed 80 years. While the majority of this longevity increase is due to declines in infant, child, and maternal mortality over the past two centuries, life expectancies have increased significantly at older ages, as well. Americans who have reached age 65 today can expect to live another 17 years or; this is 6 years longer than the predicted life span at age 65 in 1900. Due, in part, to increases in the size of the global population as a whole, many million more people than ever before are living past the age of 100 years. Some demographers argue that the absolute upper limits of human life span have not yet been reached, and that some babies born in this decade can expect to live over 130 years (Carnes et al. 2003; Vaupel 2010).

Longer life expectancies mean that we are now routinely outliving our ancestral reproductive life span. Some of our forebears, like some people living in traditional, unindustrialized cultures today, also outlived their fertile years, and older adults in both traditional and modern societies clearly can make substantial cultural contributions when they do survive into grandparenthood (Kaplan 1997; Hawkes et al. 1998; Hawkes 2004). But the current consensus among human demographers, biogerontologists, and evolutionary biologists is that life spans of over 50 years were achieved very rarely among ancient humans, and that before the industrial revolution most people lived less than 35 years—an age marked by waning fertility in modern humans, both male and female (Crews 2003; Finch and Crimmins 2004; Finch 2007; Olshansky 2014). The continuing shift toward lower mortality rates and higher life expectancies for modern people is persuasive evidence that, on average, modern people are indeed healthier, and tend to remain infection—and accident-free much longer than either our ancestors or modern counterparts still living in traditional cultures.

The recent cultural and technological innovations responsible for the modern demographic and epidemiological shift include improvements in medical care, public sanitation and safety, nutrition, and health education. By altering our environment, we have successfully altered the causes of human mortality and, in evolutionary terms, transcended the shadow of natural selection that hung over humans in centuries past. This new and growing likelihood of survival into an extended post-reproductive life span represents an unprecedented gift of modern life. But at the same time, a more "extreme" life spans means that many more of us face a much longer segment of life dealing with diseases associated with aging, including cardiovascular disease, cancers, chronic obstructive pulmonary disease, and dementias. We also confront a much higher prevalence of potentially disabling aging-related conditions, such as musculoskeletal problems (including osteoarthritis and osteoporosis), sensory deficits, and incontinence.

9.2.1 Ancestral Life spans in Perspective

From Paleolithic times until the 1800s, human populations generally experienced much higher mortality rates at all ages than those in industrialized cultures (Crews 2003; Finch and Crimmins 2004; Finch 2007), with *most* ancient people living less than 30 years. According to evolutionary theory, the force of natural selection is limited to the reproductive segment of the human life span. As fertility wanes, so does the force of selection. In a discussion of modern human demography in evolutionary perspective, Crews (2003) stated that "Australopithecines and early *Homo* species are estimated to have averaged only about 15–20 years of life over the 280,000 – 350,000 generations they prowled the earth…, while over 40 generations of early agriculturalists and nomadic pastoralists could expect to live only about 25 years…."

While some of our ancestors are likely to have lived beyond their peak reproductive years on average, and the *maximum* life span in some populations may exceeded this period, the average life expectancy for most prehistoric human populations was in all probability under 40 years (Table 9.2). According to evolutionary theory, aging is expected to occur outside the window of strong selection, after the probability of reproduction has declined significantly. Hence it is not surprising that the prevalences of aging-related diseases increase sharply after age 30, and that the slopes of increasing prevalence of many diseases and forms of disability (including cardiovascular disease, cancers, osteoarthritis, and osteoporosis) rise again after age 50, when reproduction ceases for women (Figs. 9.2, 9.3, 9.4, and 9.5).

9.2.2   The Female Longevity Paradox

Human females, like females of many other mammalian species, exhibit lower mortality rates than males at all life stages (Austad 2006, 2011). Women live significantly longer than men worldwide, and older women die at lower rates at all ages from cardiovascular diseases and cancers. Despite the fact that women are better survivors on average than men, however, older women tend to have higher rates of comorbid diseases (occurring together) and disabling conditions—including dementias, blindness, obstructive lung disease, obesity, and painful, chronic musculoskeletal problems (Figs. 9.3, 9.4, and 9.5). Since more women than men remain alive at older ages, the total population of women with potentially deadly or disabling aging-related diseases far exceeds that of men in industrialized nations (World Health Organization 2003).

The extended life spans enjoyed by humans in the twenty-first century are a product of cultural innovations, rather than rapid evolutionary changes in human biology. The reproductive segment of the human life course can be seen as the product of developmental programs and physiological trade-offs evolved to optimize reproductive success in ancestral environments. In all likelihood, however, these fitness trade-offs did not arise as a direct result of natural selection to optimize our *health* in the latter third of our lives. Our bodies were not "designed to fail"; rather, some of the health challenges now associated with aging can be understood as by-products of physiological processes optimized for reproduction, rather than for sustained health under conditions of extreme longevity.

9.3   Menopause

9.3.1   Endocrine Pleiotropy

Steroid reproductive hormones, like estrogen and testosterone, provide a particularly good illustration of how specific physiological mechanisms involved in reproductive trade-offs can be implicated in aging-related conditions. Sex steroid hormones are pleiotropic: that is, each hormone has multiple phenotypic effects (Zera et al. 2007; Bribiescas and Ellison 2008). They act during critical periods of development, coordinating physiological functions with environmental or social

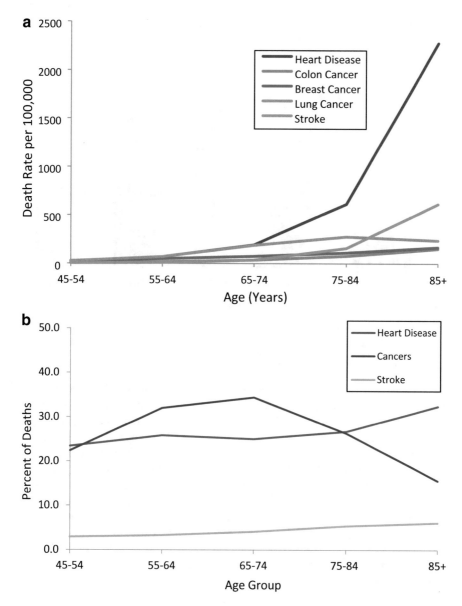

Fig. 9.2 (**a**) Death rate or (**b**) percent of deaths from several leading causes of death in older (**a**) women and (**b**) men in the USA in 2009. Mortality from heart disease and stroke for women increases at much higher rates with age than from breast, colon, or lung cancer (death rates for women expressed as crude rates per 100,000 population, all races combined). *Data source*: Centers for Disease Control and Prevention, National Center for Health Statistics, CDC WONDER Online Database, 2012

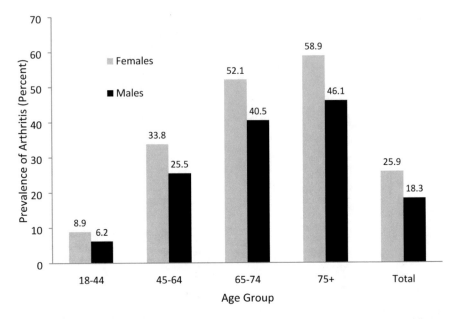

Fig. 9.3 Sex differences in prevalence of osteoporosis in adults (all races combined) at different ages in the USA, 2005–2008. Prevalence is expressed as percent of people in each group. *Data source*: National Institutes of Health (2011). Redrawn with permission

cues so as to fine-tune developmental or reproductive timing. In the context of life-history evolution, reproductive hormones are among the primary mechanisms likely to mediate fitness trade-offs shaped by natural selection. They are important for maintaining phenotypic plasticity, and provide physiological mechanisms for calibrating adaptive responses to environmental variability (Bribiescas and Ellison 2008; Gilbert and Epel 2008; Gluckman et al. 2009) (refer to introductory material on hormones, nutrition, timing of menarche, etc.). Sex hormones are associated in particular with the evolution of strategies for energy allocation and life-history trade-offs between reproduction and self-maintenance. At older ages, however, after fertility has waned, the physiological landscape of these endocrine trade-offs shifts. Hormones that were essential for reproduction—or the absence of these hormones—can have important, and sometimes paradoxical, effects on the health of older humans.

9.3.2   Human Menopause in Comparative Context

The most straightforward physiological explanation for human menopause is ovarian aging. Menopause marks the depletion of finite stores of viable oocytes, accompanied by declining levels of estrogen and progesterone followed by the cessation of ovulation and, ultimately, infertility. While humans are unusually long-lived

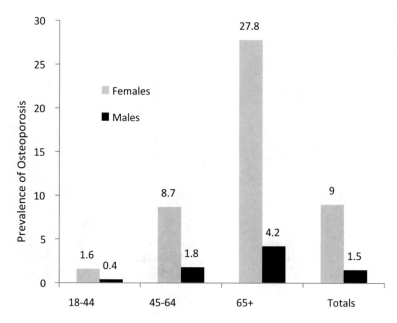

Fig. 9.4 Sex differences in prevalence of osteoarthritis in adults (all races combined) at different ages in the USA, 2007–2009. Prevalence is expressed as percent of people in each group who reported a diagnosis of osteoarthritis by a medical professional. *Data source*: National Institutes of Health (2011). Redrawn with permission

for our average body size, a life cycle marked by midlife cessation of fertility followed by a post-reproductive period of some length is not unique to humans. The underlying physiology of ovarian failure is poorly studied in many nonhuman animals (with the exception of laboratory rodents), but comparative surveys show that many other female vertebrates, including nonhuman mammals, exhibit reproductive aging patterns consistent with the depletion of ovarian reserve (for reviews, see Austad 1994; Cohen 2004; Finch and Holmes 2010). Female post-reproductive life spans of up to a third of the total life span are documented in eight separate mammalian orders, suggesting that the underlying ovarian physiology is evolutionarily conserved. These examples are not limited to species with particularly long life spans, well-developed maternal care, or extended kin networks. Post-menopausal life spans have also been documented in rodents, domestic birds (e.g., Japanese quail), and some captive wild fishes (e.g., guppies) (Holmes et al. 2003; Reznick et al. 2006; Finch and Holmes 2010). Post-reproductive life spans are more likely to be seen when animals are maintained in captivity and protected from natural sources of mortality. In nature, they appear to be much rarer, and limited to long-lived, highly social vertebrates like elephants and pilot whales.

Extended post-reproductive life spans in humans and other social mammal species with well-developed parental care and overlapping generations have been suggested to represent an adaptation for increasing females' inclusive fitness (reproductive success of close kin) and for enabling grandmothers to assist

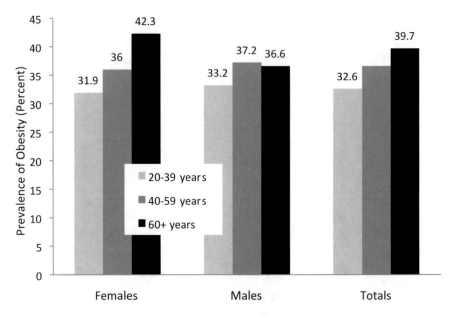

Fig. 9.5 Sex differences in prevalence of obesity in adults (all races combined) of three different age groups in the USA, 2007–2009. Prevalence is expressed as a percentage of each group. *Data source*: National Institutes of Health (2011). Redrawn with permission

with care and provisioning of grandoffspring. Healthy grandmothers (and grandfathers, as well) clearly do make social and nutritional contributions that benefit younger kin (Kaplan 1997; Hawkes et al. 1998; Hawkes 2004; reviewed in Jasienska 2013). However, *the female reproductive life span in humans, as well in other animals, likely reflects selection by ancestral mortality forces, and corresponds to the likelihood of survival in past environments.* As the likelihood of survival in modern cultures has become uncoupled from ancestral mortality risk, vast numbers of women are living much longer now than ever before in a post-reproductive state. In the face of this more extreme modern longevity, the loss of circulating sex steroids has new and significant consequences over the longer term.

Menopause is considered to be a healthy reproductive transition, rather than a pathological state (Goodman et al. 2011; Stuenkel et al. 2012). Nonetheless, many of the health issues now facing older women arguably are associated with living significantly longer than ever before in a state characterized by low—or post-reproductive levels of reproductive hormones. We need not view this newly extended segment of the life span in "either-or" terms of adaptation vs. pathology. Instead, the post-menopausal years can be seen as a "new normal"—constrained to some extent by past selection for reproductive success, but now representing a special phase of the life cycle in which longer-term health may be optimized using strategic new approaches to disease prevention and intervention, as informed by an evolutionary perspective.

9.3.3   Estrogen and Health

The concept of endocrine pleiotropy is particularly relevant to aging-related health and disease in women. Estrogens promote gamete development, as well as growth and cell proliferation in target reproductive tissues—actions that are essential for female sex differentiation and reproduction. As the ovaries regress at menopause (around age 50), estrogen and progesterone levels drop precipitously. In modern humans, the timing of menopause corresponds with changes in the slopes (rates of increase) of age-specific prevalences of many diseases—including cardiovascular disease and some cancers—as well as increased rates of mortality (Fig. 9.2).

Loss of estrogen is clearly linked to such potentially disabling, aging-related conditions as osteopenia and osteoporosis, which are characterized by depletion of calcium in bones that can result in bone fragility, frailty, falls, and fractures. Osteoporosis affects older women in far greater numbers than men (Fig. 9.3). Estrogen may, moreover, play a role in preventing osteoarthritis, another aging-related condition that disables more women than men (Fig. 9.4). Estrogen loss is also clearly responsible for troublesome symptoms many women experience during the menopausal transition and beyond (Goodman et al. 2011; Stuenkel et al. 2012). Some of these are serious enough to cause real distress and require medical intervention—including hot flashes or flushes, vulvovaginal atrophy, painful intercourse, and loss of libido. Many women experience additional symptoms that significantly impact their well-being, including anxiety, difficulty sleeping, memory problems, and irritability.

A wealth of research of various types (including clinical, epidemiological, and laboratory animal studies) suggests that, at least under certain physiological conditions, estrogen has cardioprotective, neuroprotective, anti-inflammatory, and antioxidant properties (reviewed in Brinton and Nilsen 2003; Chen et al. 2006; Moolman 2006; L'Hermite et al. 2008; Bay-Jensen et al. 2012). Estrogen loss is linked by several lines of evidence to cardiovascular disease, including heart attack and stroke. The precise nature of this link remains controversial and is currently an intense focus of research. Epidemiological studies of the effects of early (surgical or natural) menopause suggest that estrogen protects against cardiovascular disease, and decreases the overall mortality rate in older women (see, for example, Mikkola and Clarkson 2002; Rocca et al. 2006; Stuenkel 2012). One recent study of over 2500 women aged 45–84 presented evidence that a subset of about 693 woman reporting menopause (surgical or natural) at the age of 46 or younger had approximately double the risk of heart disease or stroke compared to women reaching menopause later (Wellons et al. 2012).

The "dark side" of estrogen pleiotropy involves its capacity to promote the proliferation of cells (including cancer cells) in target tissues. Estrogen is clearly implicated as a risk factor for various reproductive cancers (breast, uterine, and ovarian) in women, and prevalence rates of all reproductive cancers increase with age (American Cancer Society 2013) (Fig. 9.2a). (Although men can get breast cancer, and it can be deadly in men, it is predominantly a women's disease.) Interestingly, however, rates of cardiovascular disease and dementia increase much more dramatically for women after menopause, suggesting a protective role for estrogen. Current research suggests that there is a "window of opportunity" for the therapeutic effects of estrogen therapy, however, with past exposure to endogenous levels of this hormone being a key predictor of therapeutic action, as well as the risks of cardiovascular disease and reproductive cancers later in life. This therapeutic window is reminiscent of the critical periods for hormone action associated with milestones during normal development and the reproductive life span.

Some very large and well-publicized controlled clinical trials, including the Women's Health Initiative (WHI) study in the USA, have shown an increased risk of breast cancer and stroke in women who started hormone therapy (estrogen combined with progestogen) late (in the late 1950s or older, well after the menopausal transition), and continued it for an extended period of time (Roussouw et al. 2007; Harman et al. 2011). In the WHI study, however, as well as some later clinical trials, hormone therapy has been shown to pose very little health risk for younger menopausal women (starting around age 50) when used to treat troublesome symptoms for five or more years during and after the transition. These findings reinforce the idea that a single hormone may have pleiotropic effects—some good and some bad—and that during reproductive aging, as during earlier development, the timing of hormone action is critical. Researchers attempting to develop strategies for safe treatment of menopausal symptoms are currently conceptualizing an optimal "window of opportunity" during the reproductive aging process intended to balance benefits and risks of hormone therapy. While estrogen currently is not approved for the prevention of aging-related diseases except osteoporosis (i.e., low-dose transdermal patch), the physiological complexity of the role that this hormone plays in cardiovascular and neurological disease, including dementia, is still a focus of intense study and debate. A number of prominent clinician–researchers have argued for a greater role of estrogen therapy in the primary prevention of aging-related disease (Langer et al. 2012; L'Hermite 2013).

The arc of life has shifted dramatically for modern humans. As the baby boomers approach old age, we are participating in a demographic and epidemiological revolution. The extension of longevity far beyond the fertile life span, while not a common part of our ancestors' life history, has become the "new natural."

We are living longer, largely healthier lives in industrialized countries because we have been released from the selection pressure and sources of mortality that our ancestors experienced. Viewed in this evolutionary context, and in terms of the relative recency of historical changes in the human life span, it is not surprising that we face new challenges for optimizing health during the latter third of life. Millions of us are living outside the "selection shadow" (Kirkwood and Austad 2000) that shaped our ancestral life span, and we are now faced with "new" (in terms of increased incidence in the population) set of chronic diseases and disabilities associated with modern life.

The benefits and blessings of living longer, healthier lives are obvious. The "down side" of living to older ages is that we are now surviving long enough to contract diseases of aging that seldom affected humans in past eras, including cancers, diabetes, dementias, cardiovascular disease, and chronic obstructive pulmonary disease. In addition, we are encountering in the modern environment new physiological challenges, with which we may be rather poorly designed to cope for the long haul (Finch and Crimmins 2004; Finch 2007; Gurven et al. 2008). These include extended periods of exposure to physiological stressors, toxins, and potentially damaging metabolic by-products, including inflammatory compounds and products of oxidative metabolism implicated in aging and disease (reviewed in Holmes and Cohen 2014). The negative effects of modern overnutrition and a sedentary lifestyle, moreover, may combine with those of an extended post-reproductive life span, in which levels of sex hormones are very low, producing higher rates of potentially disabling conditions in older women like osteoporosis, osteoarthritis, diabetes, and obesity.

9.3.4 Surviving into the Gray Zone

Living in this evolutionary "gray zone" of extreme longevity poses some new (and sometimes paradoxical) health challenges. Menopause, with its associated changes in levels of estrogen and progesterone, presents a case in point that is challenging in its physiological and cultural complexity. Extended post-menopausal life spans may not have been a "natural" feature of our ancestors' life histories, but they are now normal for millions of women in industrialized countries (see, for example, Pollycove et al. 2011). Many otherwise healthy women experience—and request medical support for—challenging symptoms during and following the menopausal transition. Although these symptoms vary a great deal among modern cultures, in the USA and Europe they are problematic for many women. In the USA and Europe, clinical trials and reliable survey data show that over 20 % of women experience menopausal symptoms troublesome enough to significantly affect their overall well-being, including vasomotor symptoms (hot flashes or flushes), cognitive issues, anxiety, irritability, and changes in sexual functioning (Sowers et al. 2013; L'Hermite 2014). Up to 40 % of women, moreover, experience less severe symptoms. As the cultural context for menopause has shifted along with the life course, expectations for health and well-being for older women have changed as well.

9.3.5 Windows of Opportunity

The health effects of steroid hormones during aging, as during reproductive life, are dependent on life stage and physiological context. Their effects on older women are also pleiotropic, with multiple, complex impacts on health, disease risk, and longevity. Like many other aspects of aging, the effects of sex steroids are unpredictable, individually variable, and difficult to place into an adaptive context, in which physiological phenomena are expected to have a phenotypically adaptive range of effects.

As the life spans of modern women have lengthened, weighing the medical trade-offs associated with estrogen therapy during the menopausal transition and beyond has become a major challenge for medical practitioners and the research community concerned with older women's health.

Post-menopausal women are at increasingly greater risk of breast cancer as they age, and this risk can be increased by the use of menopausal hormone therapy. The risks of death and disability from heart disease and stroke, however, are much higher, and increase with age at a much higher rate in older women (Fig. 9.2a). Estrogen may afford protection against cardiovascular disease. Current thinking in the clinical menopause research community is that there are critical windows of time for treating menopausal symptoms that can afford an acceptable balance between benefits and risks. The concept of *critical clinical windows* echoes that of critical periods of responsiveness to endogenous hormones during development. During reproductive aging, however, these windows are dissociated from their earlier developmental and reproductive contexts, in which the maximization of adaptive reproductive trade-offs is paramount—sometimes at the expense of long-term health and survival.

Women living into the age of grandmothering undoubtedly make social, economic, and other contributions to their extended social networks, regardless of the culture in which they find themselves. On the other hand, most modern women are delaying or significantly reducing their reproductive output, or have foregone reproduction altogether. Ideally, health priorities in the new millennium take into account these changes in the arc of women's lives, and focus on maximizing healthy life span and physical, social, and psychological well-being for women, while recognizing that we are not entirely free of ancestral physiological and evolutionary constraints.

References

Ackermann M, Pletcher S (2008) In: Stearns SC, Koella J (eds) Evolutionary thinking as a foundation for studying aging and aging-related disease. Oxford University Press, Oxford

American Cancer Society (2013) Cancer facts and figures 2013. American Cancer Society, Atlanta

Appleby A (1980) Epidemics and famine in the little ice age. J Interdiscip Hist 10:643–663

Austad SN (1994) Menopause: an evolutionary perspective. Exp Gerontol 29:255–263

Austad SN (2011) Sex differences in longevity and aging. In: Masoro EJ, Austad SN (eds) Handbook of the biology of aging. Academic, New York

Austad SN (2006) Why women live longer than men. Gend Med 3:79–92

Bay-Jensen AC, Slagboom E, Chen-An P, Alexanderson P, Christiansen C, Meulenbelt I, Karsdal M (2012) Role of hormones in cartilage and joint metabolism: understanding an unhealthy metabolic phenotypic in osteoarthritis. Menopause 20:578–586

Bell FC, Miller SI (2005) Life tables for the United States Social Security area 1900–2100. U.S. Social Security Administration, Report No. SSA, Pub. No. 11–11536

Bribiescas R, Ellison PT (2008) How hormones mediate trade-offs in human health and disease. In: Stearns SC, Koella J (eds) Evolution in health and disease. Oxford University Press, New York

Brinton RD, Nilsen J (2003) Effects of estrogen plus progestin on risk of dementia. JAMA 290 :1706–1708

Carnes BA, Olshansky SJ, Grahn D (2003) Biological evidence for limits to the duration of life. Biogerontology 4:31–45

Chen SH, Nilsen J, Brinton RD (2006) Dose and temporal pattern of estrogen exposure determines neuroprotective outcome in hippocampal neurons: therapeutic implications. Endocrinology 147:5303–5313

Cohen AA (2004) Female post-reproductive lifespan: a general mammalian trait. Biol Rev Camb Philos Soc 79:733–750

Cohen AA, Holmes DJ (2014) Evolution and the biology of aging. Reference module in Biomedical Sciences. Elsevier. 15-Oct-14. doi:10.1016/B978-0-12-801238-3.00032-5

Crews D (2003) Human senescence: evolutionary and biocultural perspectives. Cambridge University Press, Cambridge

Finch C, Holmes D (2010) Ovarian aging in developmental and evolutionary contexts. Ann N Y Acad Sci 1204:82–94

Finch CE (2007) The biology of human longevity: inflammation, nutrition, and aging in the evolution of lifespans. Academic/Elsevier, Amsterdam

Finch CE, Austad S (2008) The evolutionary context of human aging and degenerative disease. In: Stearns SC, Koella JC (eds) Evolution in health and disease. Oxford University Press, New York

Finch CE, Crimmins EM (2004) Inflammatory exposure and historical changes in human life-spans. Science 305:1736–1739

Finch CE, Kirkwood TBL (2000) Chance, development, and aging. Oxford University Press, New York

Gilbert S, Epel D (2008) Ecological developmental biology. Sinauer Associates, Sunderland

Gluckman PD, Hanson MA, Bateson P, Beedle AS, Law CM, Bhutta ZA, Anokhin KV, Bougneres P, Chandak GR, Dasgupta P, Smith GD, Ellison PT, Forrester TE, Gilbert SF, Jablonka E, Kaplan H, Prentice AM, Simpson SJ, Uauy R, West-Eberhard MJ (2009) Towards a new developmental synthesis: adaptive developmental plasticity and human disease. Lancet 373:1654 –1657

Guttormsson L, Garðarsdóttir Ó (2002) The development of infant mortality in Iceland, 1800–1920. Hygiea Internationalis 2:151–176

Goodman N, Cobin R, Ginzburg S, Katz I, Woode D (2011) American Association of Clinical Endocrinologists medical guidelines for clinical practice for the diagnosis and treatment of menopause. Endocr Pract 17:949–954

Gurven M, Kaplan H, Winking J, Eid Rodriguez D, Vasunilashorn S, Kim JK, Finch C, Crimmins E (2009) Inflammation and infection do not promote arterial aging and cardiovascular disease risk factors among lean horticulturalists. Plos One 4, e6590

Gurven M, Kaplan H, Winking J, Finch C, Crimmins EM (2008) Aging and inflammation in two epidemiological worlds. J Gerontol A Biol Sci Med Sci 63:196–199

Harman SM, Vittinghoff E, Brinton E, Budoff M, Cedars M, Lobo R et al (2011) Timing and duration of menopausal hormone treatment may affect cardiovascular outcomes. Am J Med 124:199–205

Hawkes K (2004) Human longevity: the grandmother effect. Nature 428:128–129

Hawkes K, O'Connell JF, Jones NGB, Alvarez H, Charnov EL (1998) Grandmothering, menopause, and the evolution of human life histories. Proc Natl Acad Sci U S A 95:1336–1339

Holmes DJ, Cohen AA (2014) Overview: aging and gerontology. Reference module in biomedical sciences. Elsevier, Amsterdam. doi:10.1016/B978-0-12-801238-3.00149-5

Holmes DJ, Thomson SL, Wu J, Ottinger MA (2003) Reproductive aging in female birds. Exp Gerontol 38:751–756

Jasienska G (2013) The fragile wisdom: an evolutionary view on women's biology and health. Harvard University Press, Cambridge

Kaplan H (1997) The evolution of the human life course. In: Wachter KW (ed) Between Zeus and the Salmon. National Academy Press, Washington

Kennedy BK, Berger SL, Brunet A, Campisi J, Cuervo AM, Epel ES, Franceschi C, Lithgow GJ, Morimoto RI, Pessin JE, Rando TA, Richardson A, Schadt EE, Wyss-Coray T, Sierra F (2014) Geroscience: linking aging to chronic disease. Cell 159:709–713

Kirkwood TB, Austad SN (2000) Why do we age? Nature 408:233–238

Kirkwood TBL (1981) Repair and its evolution: survival vs. reproduction. In: Townsend CR, Calow P (eds) Physiological ecology: an evolutionary approach to resource use. Sinauer Associates Inc, Sunderland

Langer RD, Manson JE, Allison MA (2012) Have we come full circle—or moved forward? The Women's Health Initiative 10 years on. Climacteric 15:206–212

L'Hermite M (2013) HRT optimization, using transdermal estradiol plus micronized progesterone, a safer HRT. Climacteric Supplement 1:44–53

L'Hermite M (2014) Aging: menopause and hormone treatment. In: Reference module in biomedical sciences. Elsevier, Amsterdam

L'Hermite M, Simoncini T, Fuller S, Genazzani AR (2008) Could transdermal estradiol plus progesterone be a safer postmenopausal HRT? A review. Maturitas 60:185–201

Lithgow GJ (2013) Origins of geroscience. Public Policy Aging Rep 23:10–11

Manton KG, Gu X, Lamb VL (2006) Long-term trends in life expectancy and active life expectancy in the United States. Popul Dev Rev 32:81–105

McDonald RB (2014) Biology of aging. Garland science. Taylor & Francis Group, New York

Mikkola TS, Clarkson TB (2002) Estrogen replacement therapy, atherosclerosis and vascular function. Cardiovasc Res 53:605–619

Moolman J (2006) Unravelling the cardioprotective mechanism of action of estrogens. Cardiovasc Res 69:777–780

National Center for Health Statistics (2012) Underlying cause of death 1999–2009. CDC Wonder. Data Compiled from the Multiple Cause of Death File 2009, Series 20 No. 2O, 2012

Olshansky SJ (1992) Estimating the upper limits to human longevity. Popul Today 20:6–8

Olshansky SJ (2014) Demography of human aging and longevity. Reference module in biomedical research, 3rd edn. doi:10.1016/B978-0-12-801238-3.00150-1

Olshansky SJ, Carnes BA, Cassel C (1990) In search of Methuselah: estimating the upper limits to human longevity. Science 250:634–640

World Health Organization (2003) Gender, health and ageing. Department of Gender and Women's Health, World Health Organization, Geneva

Partridge L, Barton NH (1993) Optimality, mutation and the evolution of ageing. Nature 362 :305–311

Pollycove R, Naftolin F, Simon JA (2011) The evolutionary origin and significance of menopause. Menopause 18:336–342

Promislow DEL, Fedorka KM, Burger JMS (2006) Evolutionary biology of aging: future directions. In: Masoro EJ, Austad SN (eds) Handbook of the biology of aging. Academic, New York

Reznick D, Bryant M, Holmes D (2006) The evolution of senescence and post-reproductive lifespan in guppies (*Poecilia reticulata*). PLoS Biol 4

Rocca WA, Grossardt BR, de Andrade M, Malkasian GD, Melton J (2006) Survival patterns after oophorectomy in premenopausal women: a population-based cohort study. Lancet 7:821–828

Rose MR (1991) Evolutionary biology of aging. Oxford University Press, New York

Roussouw J, Prentice R, Manson J et al (2007) Effects of postmenopausal hormone therapy on cardiovascular disease by age and years since menopause. JAMA 297:1465–1477

Sowers M, Harlow SD, Karvonen C, Bromberger J, Cauley J, Gold E, Matthews K (2013) Menopause: its epidemiology. In: Goldman MB, Troisi R, Rexrode K (eds) Women and health. Academic, New York

Stearns SC (1992) The evolution of life histories. Oxford University Press, New York

Stuenkel C (2012) Cardiovascular risk and early menopause: cause or consequence?

Stuenkel, C., Gass, M., Manson, J., Lobo, R., Pal, L., Rebar, R., & Hall, J. (2012). A decade after the Women's Health Initiative - the experts do agree. *Menopause, 19(8)*

Tatar M, Bartke A, Antebi A (2003) The endocrine regulation of aging by insulin-like signals. Science 299:1346–1351

Vaupel JW (2010) Biodemography of human ageing. Nature 464:536–542

Wellons M, Ouyang P, Schreiner P, Herrington D, Vaidya D (2012) Early menopause predicts future coronary heart disease and stroke: the multi-ethnic study of atherosclerosis. Menopause 19:1081–1087

Williams GC (1966) Adaptation and natural selection. Princeton University Press, Princeton

Williams GC (1957) Pleiotropy, natural selection and the evolution of senescence. Evolution 11:398–411

Zera A, Harshman L, Williams TD (2007) Evolutionary endocrinology: the developing synthesis between endocrinology and evolutionary genetics. Annu Rev Ecol Evol Syst 38:793–817

Zera AJ, Harshman LG (2001) The physiology of life history trade-offs in animals. Annu Rev Ecol Evol Syst 32:95–126

Chapter 10
Costs of Reproduction, Health, and Life Span in Women

Grazyna Jasienska

Abstract Reproduction in females is costly: pregnancy, breast-feeding, and child-care require energy, and thus energetic costs are indispensable feature of reproduction. Pregnancy and lactation also require many physiological changes, including in maternal immune system and increased levels of oxidative stress. Finally, genes involved in encoding traits related to reproduction often have multiple functions. Some of these genes have alleles that support reproduction but also increase risk of diseases later in life, and their carriers have higher mortality.

Results of studies testing relationships between reproduction and health and reproduction and life span are contradictory, possibly due to methodological problems and theoretical framework problems. There is a need for studies that would analyze these relationships at both genetic and phenotypic levels, and that would comprehensively calculate costs of reproductive investment, including not only number of children but also birth spacing, lactation, childcare, and extended reproductive effort.

10.1 The Principle of Allocation

Parental investment is costly and costs of reproduction are higher for human females than for males due to energetic and nutritional requirements of pregnancy and lactation, and traditionally female-dependent childcare. Energetic costs of pregnancy result from fetal growth, growth and maintenance of maternal supporting tissues, and maternal fat accumulation (Blackburn and Loper 1992). Energetic demands of lactation are a consequence of milk synthesis and the maintenance of metabolically active mammary glands (Lunn 1994). Human children require support also after weaning and in traditional societies became independent at about 15 years of age (Kramer 2005).

G. Jasienska (✉)
Department of Environmental Health, Jagiellonian University Collegium Medicum,
Grzegorzecka 20, 31-531 Krakow, Poland
e-mail: jasienska@post.harvard.edu

© Springer Science+Business Media New York 2017
G. Jasienska et al. (eds.), *The Arc of Life*, DOI 10.1007/978-1-4939-4038-7_10

If resources are limited, allocation to one function can only occur at the expense of other functions (Zera and Harshman 2001), and competition for common resources results in trade-offs among life history traits. Increased allocation to current reproduction should result in reduced future reproduction and/or reduced life span. Pregnancy and lactation take energy and nutrients away from other processes and cause multiple metabolic changes. Thus it is expected that female physiology will be negatively affected by the reproductive effort. Repeated reproductive events may negatively affect maternal health, especially at older age, and, ultimately, increase risk of mortality.

10.2 Energetic Costs of Menstrual Cycle and Pregnancy

Ovarian and uterine function in adult females must be constantly maintained. This maintenance requires energy but is not very expensive. During several days of the luteal phase of the cycle, an increase of 6–12 % in the resting metabolic rate has been reported in some women (Bisdee et al. 1989; Curtis et al. 1996; Meijer et al. 1992). An average basal metabolic rate (BMR) for a woman from an industrialized country is 1300 kcal/day (Leonard 2008); thus the increase due to menstrual cycle is between 70 and 150 kcal/day. Some additional energy beyond the regular metabolic needs is required to support regular menstrual function (Strassmann 1996a, b); these costs, however, are negligible in comparison to the much higher energetic demands of pregnancy and lactation.

For well-nourished women from industrialized countries, the estimated costs of pregnancy constitute an expense of an extra 90, 290, and 470 kcal/day, respectively, for the first, second, and third trimesters (Butte and King 2005). During the last trimester, the woman may require up to 22 % of additional energy over the prepregnant values (Butte et al. 1999).

10.3 Costs of Lactation and Childcare

The cost of lactation is a very important variable in calculations of total reproductive costs because, on average, 1 day of lactation places higher demands on maternal energetics than 1 day of pregnancy. Costs of lactation change with the age of the infant (Rashid and Ulijaszek 1999) and the frequency of daily feedings, but, on average, lactation requires additional 626 kcal/day (Butte and King 2005) and may last for a few years. Lifetime costs of lactation are usually neglected by research analyzing relationships between fertility and longevity of women. Data on duration of breast-feeding for individual women are not available from historical demographic records, and omissions of these costs may significantly bias the results of research on reproduction and longevity. Errors may be substantial when women from different socioeconomic classes are compared, since it is likely that there was a considerable variation in breast-feeding practices among such groups.

In addition to costs of pregnancy and lactation, there are also often significant costs which are associated with childcare or with an increase in intensity and duration of work necessary to obtain resources to support each additional child (Sujatha et al. 2000). Estimates of these costs are hard to find as well. Women who do not have any additional help clearly spend more time and energy as the number of children increases. However, as children grow older, they are able to provide childcare to younger siblings and become involved in housework, agricultural work, or work for wages (Biran et al. 2004; Kramer 2002; Nag et al. 1978). Therefore, the relationship between the number of children and the maternal costs of childcare is clearly not straightforward.

10.4   The Intergenerational Costs: Children and Grandchildren

Children not only impose biological costs but also provide some benefits to their parents, for example, by contributing labor to the household. In preindustrial societies, characterized by the absence of pension system or health insurance, having children was especially important for elderly parents (Tsuya et al. 2004). How children affect parental well-being depends on economic conditions and social structure of the population, especially family structure, and behavioral patterns of taking care of aging parents. For example, in the Liaoning Province in China between 1749 and 1909, women had 12 % higher mortality if they did not have any living sons, and the risk of mortality for women without sons increased further if women were widows (Campbell and Lee 2004). Having children had a positive effect on parental life expectancy also in some historical European communities (Tsuya et al. 2004).

Parental investment may continue for many years, and once parents complete their direct investment to children, they often became involved in helping to raise grandchildren. This prolonged investment requires additional energy and as such may increase burden associated with reproductive costs; thus well-being of parents is not only influenced by costs and benefits resulting from interactions with their children but, in later years, also with their grandchildren. Older people lived and, in some societies still do, with one of their children and his or her offspring. Costs and benefits of such interactions are important to consider when discussing health and mortality risks in relation to reproductive investment. While the impacts of grandparents on health and survival of grandchildren are widely studied (Sear et al. 2000; Hawkes et al. 1997; Voland and Beise 2002), less interest has been devoted to whether these interactions have any effect on the well-being of grandparents themselves. In patrilineal systems, cost and benefits of interactions with grandchildren may be different for grandmothers and grandfathers. Expectations toward grandmothers may be much higher than for grandfathers and may include childcare or physical labor (Pashos 2000). Grandfathers may feel entitled to a higher share of resources than those left for grandmothers which may be especially important when resources are scarce, and some studies report higher incidence of malnutrition in older females than in older males (Aliabadi et al. 2008). Even in modern families, grandmothers are expected to provide more care and support to families, as shown, for example, by a study investigating roles of grandparents in support networks in neonatal intensive care units in Scotland (McHaffie 1992).

Costs related to interactions with grandchildren may vary for maternal and paternal grandmothers. Maternal grandmothers, due to paternity uncertainty (Hamilton 1964), are expected to contribute more toward children of their daughters than to children of their sons. This prediction has been confirmed by many studies documenting that maternal grandmothers are more often the primary caregivers (Euler and Weitzel 1996; Michalski and Shackelford 2005; Sear and Mace 2008). Several studies documented benefits for grandchildren living with maternal grandmothers and lack of such benefits when children live with their paternal grandmothers (Hawkes et al. 1997, 1998; Jamison et al. 2002; Sear et al. 2000; Voland and Beise 2002). It is very likely that providing such benefits for children is not cost-free for their grandmothers. Childcare and resource acquisition are energetically demanding and may negatively affect nutritional status of grandmothers.

10.5   Reproductive Costs and the Overall Energy Budgets

Female zebra finches that are experimentally forced to have high costs of reproduction by laying more eggs do not seem to suffer negative metabolic consequences, provided that their diets are supplemented (Bertrand et al. 2006). It is not clear if similar relations should be expected for human females, but it is likely that high costs of reproduction will not have the same effects on women who have good diets and low levels of physical activity as on women in poor energetic condition.

Basal metabolism, measured by basal metabolic rate (BMR), is expected to rise in women who are pregnant or nursing a child. The BMR of a pregnant woman should increase during each of the four quarters of pregnancy, respectively, by about 3, 7, 11, and 17 % above the prepregnant values (Hytten and Leitch 1971). Empirical data usually support these theoretical predictions but only in women who have good nutritional status (Butte et al. 1999; Durnin 1991, 1993; Lunn 1994; Prentice and Prentice 1990; Prentice and Whitehead 1987). However, in limited energy environments, the physiology of women who are already pregnant or lactating may be forced to rely on energy-saving strategies in order to support energetic demands of the fetus or infant (Peacock 1991). Reallocating more energy to reproduction could be achieved by reduction in the maternal basal metabolism (Peacock 1991; Prentice et al. 1989, 1995).

In pregnant women from Scotland and Gambia, who are in poor nutritional condition, the BMR showed a significant decrease up to the 12th week of pregnancy. Later BMR increased and approached the prepregnant values by 22–26 weeks and continued to rise further, but even at delivery it was still much lower than the BMR of well-nourished Swedish women (Prentice and Whitehead 1987). Pregnant Nigerian women showed significant variability in BMR, which corresponded to variability in their nutritional status (Cole et al. 1989). Even well-nourished Western women show significant variation in the BMR responses to pregnancy paralleling differences in their prepregnant body fat (Prentice et al. 1989), and maternal adipose reserves serve as highly significant predictors of changes in the BMR (Bronstein et al. 1996). Reduction in the BMR at the beginning of pregnancy seems to be an important strategy for women in poor nutritional condition. Lowering the BMR considerably reduces energetic costs of pregnancy and allows women to allocate some energy into fat storage which may be critical for the energetic support of lactation.

Lactation is predicted to cause, on average, a 12 % increase in the BMR above the nonpregnant values (Hytten and Leitch 1971); however, during lactation, the BMR has been observed to increase, decrease, or remain at prepregnant levels (Forsum et al. 1992; Goldberg et al. 1991; Guillermo-Tuazon et al. 1992; Lawrence and Whitehead 1988; Madhavapeddi and Rao 1992; Piers et al. 1995). Variation in the BMR during lactation can be explained by differences in the nutritional status of the women (Prentice and Prentice 1990). Women from the Gambia during the first year of lactation showed a 5 % decrease in the BMR compared to the prepregnant values which saves about 200 kcal/day (Lawrence and Whitehead 1988). In comparison to women who increased their BMR during lactation, savings add up to about 380 kcal/day. The saved energy can be allocated into milk production. The savings from the reduction of the BMR, although not sufficient to cover the whole expense of lactation, are still substantial, since average daily cost of milk production for women in developing countries is about 480 kcal (Lunn 1994).

Reduction in basal metabolism suggests that some of the metabolic processes (e.g., components of the maintenance metabolism, like protein turnover, or immune function) are temporarily slowed down or even halted (King et al. 1994; Prentice and Whitehead 1987); thus a long-term reduction in the BMR may be detrimental to the maternal health.

10.6   Reproduction and Life Span

The question about the impact of reproduction on life span has been addressed by many epidemiological and historical demography studies (for review see Le Bourg 2007). Several studies described a U-shaped relationship between the number of children and risk of mortality (Green et al. 1988; Lund et al. 1990; Manor et al. 2000). In the historical Swedish population (1766–1885), the number of children ever born by a woman had a negative impact on maternal longevity (Dribe 2004). Giving birth to four or more children increased maternal mortality by 30–50 % in comparison to women with fewer children. Having four or five children, instead of none or one, shortened the woman's life span by 3.5 years. However, the negative impact of parity was restricted only to landless women, while those with better socioeconomic status were not affected by reproductive costs at all.

In a historical population of northwest Germany (1720–1870), the number of children had a negative impact on longevity but only for the group of poor, landless women (Lycett et al. 2000). Among other women, with higher economic status, a *positive* relationship between the number of children and longevity was described suggesting that trade-offs between costs of reproduction and longevity apply only for those women for whom such costs are a substantial part of their overall energy budget.

Gender of children may also be an important variable to consider when analyzing the costs of reproduction and longevity. Boys have faster rate of intrauterine growth and heavier average size at birth (Loos et al. 2001; Marsal et al. 1996) and, given larger body size, perhaps higher lactational demands (Hinde 2007). Women have longer interbirth intervals after giving birth to a son than after giving birth to a daughter (Mace and Sear 1997), and offspring born after brothers have lower birth weight (Cote et al. 2003), lower height at adulthood (Rickard 2008), and also lower number of surviving children (Rickard et al. 2007) than those born after sisters (but see Puskarczyk et al. 2015). This suggests that having sons may be more energetically expensive for mothers than having daughters and that the maternal organism may become more depleted by producing male offspring.

In Finnish Sami (Helle et al. 2002) and a Flemish village (Van de Putte et al. 2003), sons decreased maternal life span, while daughters did not. In four Polish small agricultural villages, however, analyses of parish records from 1886 to 2002 showed that both the number of sons and the number of daughters decreased maternal life span and did so to the same degree (Jasienska et al. 2006a).

Not all studies documented detrimental effect of reproductive investment on maternal life span, and, in fact, a number of studies of historical populations documented a positive association between the number of children and life span, as, for example, in the Amish population (McArdle et al. 2006) and German women of higher socioeconomic status (Lycett et al. 2000). In a French-Canadian cohort of women (seventeenth and eighteenth centuries), longevity increased with increasing number of children, especially for women with the late age at last birth (Muller et al. 2002), and this may suggest slower rate of ovarian and overall aging in some women (Dribe 2004). Others did not find either positive or negative significant relationships between the number of children and life span of mothers (e.g., Le Bourg et al. 1993).

10.7 Reproduction and Health: Obesity, Diabetes, Cardiovascular Diseases, and Osteoporosis

Reproduction requires not only additional energy but also nutrients. In addition, physiological and metabolic adjustments (e.g., immunological and increased oxidative stress) associated with pregnancy may cause permanent changes in the maternal organism, especially when pregnancies are numerous. Self-reported health status, a reliable predictor of mortality (Idler and Benyamini 1997), is lower for women with at least three pregnancies and especially for women with six or more pregnancies (Kington et al. 1997). Parity is positively related to the risk of obesity, impaired glucose tolerance, non-insulin-dependent diabetes, and cardiovascular diseases.

10.7.1 Cardiovascular Diseases, Diabetes, and Obesity

Many studies suggested that high parity is related to increased risk of cardiovascular diseases. The longitudinal Framingham Heart Study and the National Health and Nutrition Examination Survey documented a positive relationship between the number of pregnancies and the subsequent development of cardiovascular disease (Ness et al. 1993). In British women with at least two children, each additional child increased the risk of coronary heart disease by 30 % (Lawlor et al. 2003). In a population-based cohort Swedish study, women with five children had about 50 % higher risk of cardiovascular disease than women with two children (Parikh et al. 2010). Having gone through six or more pregnancies increased the woman's risk of all types of strokes by 70 % (Qureshi et al. 1997).

The risk of diabetes may increase with parity as well. Among rural Australian women, those with five or more children had 28 % higher risk of diabetes than women with three or four children and 35 % higher risk than women with one or two children (Simmons et al. 2006). In Finnish women, parity of five or higher was related to 42 % higher risk of diabetes compared to the average risk experienced by women in this population (Hinkula et al. 2006). In Brussels, both parity and early age at first birth were associated with diabetes-related mortality in women (Vandenheede et al. 2012).

Pregnancy is a risk factor for obesity, and postpartum weight retention occurs in 60–80 % of women (Martin et al. 2014). In a study of the US women, each birth was associated with a 0.55 kg of permanent increase in body weight (Brown et al. 1992), while other studies reported that maternal body weight increased by 0.4–3.0 kg after each pregnancy (Harris and Ellison 1997). Among women with parity of three or more, a higher proportion was overweight than in women with lower parity. In women from Utah, a dose–response relationship was observed between the number of children and risk of obesity (Bastian et al. 2005): each additional live birth increased the risk of obesity by 11 %. In a study based on data from 65 countries, sustained breast-feeding contributed to reduction of postpartum BMI but not among relatively wealthy women (Hruschka and Hagaman 2015).

In women from developing countries, however, repeated reproductive events cause a reduction of body weight. The "maternal depletion syndrome" refers to the long-term negative changes in the maternal nutritional status, as opposed to the short-term changes associated with a single pregnancy or breast-feeding (Winkvist et al. 1992). In Papua New Guinea, the nutritional status of women decreased with parity (Garner et al. 1994). This maternal depletion occurred even though birth intervals were relatively long (3 years on average) in this population. In Turkana from northwestern Kenya, women from both nomadic and settled populations had parity-related decline in fat reserves (Little et al. 1992). In another African population, the !Kung San, the higher number of surviving children was related to lower body weight in women, while in men those with more surviving children had higher body weight (Kirchengast 2000).

Improvement in socioeconomic status seems to increase the maternal ability to resist the stress of repeated reproductive events. In Papua New Guinea, the decline in nutritional status was substantial in women who were foraging horticulturalists, while such changes were not observed among wage earners (Tracer 1991). Toba women from Argentina, a well-nourished population undergoing a transition from semi-nomadic hunter-gatherer to a sedentary, peri-urban lifestyle, did not lose excess weight gained during pregnancy despite prolonged and intense breast-feeding (Valeggia and Ellison 2003).

10.7.2   Bone Density and Osteoporosis

During pregnancy and lactation, high levels of calcium are required to support the child's developing skeleton. Such high calcium requirements are often met by mobilization of calcium from the skeleton of the mother (Prentice 2000); thus it can be hypothesized that women who had high number of pregnancies and breast-fed their children would have lower bone mineral density and therefore higher risk of osteoporosis. Each pregnancy causes 3–4.5 % decrease of bone mineral density in the lumbar region (Black et al. 2000; Drinkwater and Chestnut 1991). Bone mineral density is regarded as reliable predictor of bone strength (Karlsson et al. 2005), and a decrease of about 10 % of initial density doubles the risk of fractures in women (Cummings et al. 1995). Lumbar bone density decreases further by 3–6 % during lactation (Karlsson et al. 2001; Laskey and Prentice 1997). Changes in bone density occurring during pregnancy and lactation are reversible (Karlsson et al. 2005; Laskey and Prentice 1997), at least in women from industrialized populations, and there seems to be no consistent relationship between the number of children and bone mineral density (Bererhi et al. 1996). The absence of such relationships may be due to a rather low fertility of women from industrialized countries. A large study in Italy, with over 40,000 participants, did not show any effects of having children on bone density, but it compared women who never had a full-term pregnancy to women who only had one or two children (De Aloysio et al. 2002). Among Tsimani women, forager–farmers of lowland Bolivia, higher costs of reproduction (higher parity, shorter interbirth intervals, and earlier age at first birth) were associated with reduced bone mineral density (Stieglitz et al. 2015).

The relation between parity and bone health could be confounded by lifetime estrogen status of women. Estrogen has positive effect on bones, and estrogen deficiency plays a crucial role in the development of osteoporosis (Raisz 2005); thus the number of menstrual cycles during which estrogen is produced should be positively related to bone mineral density (Somner et al. 2004). Early menarche and late menopause increase the total time during which women can become pregnant. Bone mineral density is positively correlated with the length of reproductive span. In natural fertility populations, women who have high parity also have the longest reproductive life spans.

In women who are in good nutritional condition, a negative effect of high parity on bone density may be not detectable, since it is counterbalanced by the effect of having many high-estrogen cycles which are beneficial for bone health. The protective effect of parity on hip bone mineral density was shown, for example, in a study on Amish women with high parity (7.6 live births on average) (Streeten et al. 2005). In these women parity correlated positively with the later age at menopause and higher cumulative estrogen exposure (calculated as the age at menopause minus the age at menarche). In addition, parity increase body mass and overweight women have more adipose tissue, which produces aromatase, an enzyme converting androstenedione to estrogens. This suggests that high-parity women may have, due to greater weight gain, higher levels of estrogen in postmenopausal years (Bray 2002), which promote higher bone density (Alden 1989; Nguyen et al. 1995).

10.8   Reproduction and Risks of Reproductive Cancers

Lifetime costs of reproduction are related to an increased risk of cardiovascular disease, diabetes, and stroke even in women who are in good nutritional status. Why then studies show contradictory results on the association between reproductive investment and mortality in women? These contradictory results may be explained by the fact that the same features of reproductive life which involve the highest metabolic and physiological costs of reproduction, i.e., early first birth and high parity, may also serve a protective function, leading to decreased mortality from other diseases.

Early age at first reproduction and high number of children are the most important factors protecting women against breast cancer and other reproductive cancers (Hinkula et al. 2001; Kvale 1992; MacMahon 2006; Mettlin 1999). Breast cancer risk is also decreased by breast-feeding (Kvale 1992; MacMahon 2006), especially in women from developing countries, because feeding sessions are more frequent than in developed countries and mothers are often in relatively poor nutritional state. Frequent nursing, however, is unlikely to cause long-lasting ovarian suppression when the mother is in a good nutritional condition (Valeggia and Ellison 2004). For these reasons (low frequency of nursing and good nutritional condition), women from economically developed countries, even when they are breast-feeding for a long period of time, experience a much earlier resumption of postpartum ovarian activity.

It is possible that contradictory findings of studies on relationship between reproduction and longevity in women partially result from the fact that reproduction increases risk of some diseases (e.g., cardiovascular disease and diabetes) but reduces risk of other diseases (e.g., reproductive cancers). These trade-offs between risks of different diseases are not the same in all populations. The risk of breast cancer increases due to high, lifetime exposure to estrogens (Bernstein 2002; Jasienska et al. 2000; Key and Pike 1988). Early pregnancy, by inducing the differentiation of breast tissue, reduces its susceptibility to neoplastic transformation (i.e., development of tumors) (Balogh et al. 2006), and each subsequent reproductive event suppresses menstrual cycles. In addition, the post-pregnancy period is characterized by low levels of endogenous estrogens, and it is thought that this may further suppress potential tumor growth.

In poor agricultural societies, however, women generally have low levels of estrogen and progesterone in menstrual cycles. Low nutritional status and intense labor postpone the age of sexual maturation and in later life periodically suppress ovarian function (Jasienska 2001, 2003; Jasienska and Ellison 1998, 2004; Panter-Brick and Ellison 1994; Panter-Brick et al. 1993). Poor developmental conditions during fetal and childhood periods are also related to lower hormone levels in adult women (Jasienska et al. 2006b, c; Nunez-de la Mora et al. 2007). Therefore, women in poor populations have low lifetime exposure to estrogens. In contrast, in economically well-off populations, women have high levels of hormones in menstrual cycles and only rarely experience ovarian suppression (Ellison et al. 1993). This suggests that in well-off women, reproduction may cause reduction in risk of breast cancer, while in relatively poor women, risk of breast cancer is low in general, due to low lifetime estrogen levels. Thus trade-offs between risks of different diseases that occur in relation to reproductive investment are not the same in well-off and poor women. In both groups, women with high parity may experience increased risks of cardiovascular diseases and diabetes, but in well-off women, these elevated risks may be outweighed by significantly decreased risk of breast cancer. Thus, for well-off women with high-parity reduction in life span may not be observed.

10.9 Genetic Trade-Offs Between Reproduction and Life Span

Genes involved in encoding traits related to reproduction often have multiple functions. Some genes have alleles encoding traits that support reproduction but also increase risk of diseases later in life, and their carriers have higher mortality. Such phenomenon when a gene contributes to enhanced reproduction or survival in younger age but poorer health in older age is known as antagonistic pleiotropy. Well-studied example of a gene that shows antagonistic pleiotropy is a gene *ApoE*, encoding apolipoprotein E that is involved in cholesterol metabolism. Carriers of *ApoE4* allele have higher levels of cholesterol and, consequently, have a higher risk of hypertension and of cardiovascular and Alzheimer's diseases (Song et al. 2004). Some studies suggested that this health-detrimental allele is maintained in populations because its carriers have some benefits in younger age (e.g., better

Table 10.1 Comparative costs and benefits of reproduction and their impact on health and life span in women with high parity but differing in socioeconomic status. In well-off women, reproduction increases risks of several diseases (e.g., cardiovascular, diabetes), but these risks are offset by significant reduction in risk of breast and other reproductive cancers. In poor women, reproduction increases risks of the same diseases, but these women have low lifetime risks of breast and reproductive cancers (due to low levels of hormones in menstrual cycles), even when they are not reproducing. In addition, in poor women, energetic and nutritional expenses of reproduction cannot be easily compensated and may cause more substantial damage to the maternal organism

	Poor women	Well-off women
Energetic costs of pregnancies	High	High
Energetic costs of lactations	High	Low/none
Nutritional ability to meet costs of pregnancy and lactation	Low	High
Levels of estrogen in menstrual cycles	Low	High
Risk of maternal depletion	High	Low
Risk of obesity	Low	High
Risk of cardiovascular diseases	Increased	Increased
Risk of strokes	Increased	Increased
Risk of diabetes	Increased	Increased
Risk of osteoporosis	Not affected	Not affected
Risk of breast cancer	Slightly reduced	Greatly reduced
Risk of uterine and ovarian cancers	Slightly reduced	Greatly reduced
Life span	Reduced	Not affected/increased

development of the cognitive function (Filippini et al. 2009) or protection against infectious diseases (Oria et al. 2010). Turns out that women with *ApoE4* allele have higher levels of progesterone in their menstrual cycle which means that they have higher potential fertility (Jasienska et al. 2015). This finding suggests that the presence of selected genes may confound results of studies analyzing relationship between reproductive investment and health and life span. Women with *ApoE4* allele may have higher parity and poorer health due to the presence of this allele. Poor health in high-parity *ApoE4* carriers may have nothing to do with energetic and physiological costs of reproduction. Further, this suggests that trade-offs between reproduction and health should be studied both at phenotypic and genetic levels (Stearns 1989).

10.10   When Does Reproduction Reduce Life Span?

There is convincing evidence that reproduction is costly and related to long-term changes in female physiology; however, a negative impact of reproduction on life span is not always expected (Table 10.1). Energetic and metabolic costs of reproduction cannot be calculated just by adding up all of the woman's children (Jasienska 2013). Women with a similar number of children often differ in costs of lactation, and this difference can be very substantial. They may also differ in their ability to meet energetic costs of reproduction, due to differences in lifestyle, dietary intake, and physical activity. High costs of reproduction and its negative effects on some diseases may be outweighed by reduction in risk of other diseases. The life-style maintained by women is an important factor to consider, as for some women, intense reproduction will substantially reduce their risks of reproductive cancers, while for other women, risks of these cancers are relatively low, even when they have low parity. Indirect, extended costs of reproduction, such as those paid by grandmothers taking care of grandchildren, may also change the overall lifetime ratio of costs and benefits of reproduction.

Therefore, in women from rich environments, high parity may not translate into reduction in longevity for several reasons. Women with high parity may have rela-tively low costs of reproduction if they do not breast-feed their children. In many European populations, women who had higher socioeconomic status were not expected to nurse their children at all, and wet nurses were commonly employed. In well-off women, even if they nurse their children, the costs of pregnancy and lacta-tion are easily met by additional energy intake, especially when there are no ener-getic demands from physical activity. Furthermore, there are trade-offs between reproductive costs and benefits provided by protection against breast cancer. It is even possible that in well-off women, benefits may outweigh costs and high fertility may, in fact, be related to life span extension. Finnish grand multiparity women had higher mortality from all metabolic diseases, especially diabetes and cardiovascular diseases, than the average for the population, but these risks were clearly out-weighed by lower mortality from cancers, since their *overall* mortality was slightly lower than the population average (Hinkula et al. 2006). A negative relationship between fertility and longevity may, therefore, be expected in women who due to multiple pregnancies and breast-feeding not only have high costs of reproduction but also when these costs cannot be easily compensated by increases in dietary intake and reduction in physical activity. The most pronounced negative relation-ship between reproduction and life span should be expected when the lifestyle (poor nutrition and high work demands) of the woman leads to low lifetime estrogen lev-els, and additional reduction in these levels caused by reproductive events is rela-tively insignificant.

Study Questions

1. Why do studies on a relationship between lifetime reproductive investment and health and life span in women report contradictory results?
2. How would you design a study on a relationship between lifetime reproductive investment and health and life span in women? What group/groups of women would you study? What kind of data would you collect?

Acknowledgment This chapter was supported by grants from the National Science Centre (grant no. N N404 273440), Ministry of Science and Higher Education (grant no. IdP2011 000161) and grant K/ZDS/006113, *Salus Publica* Foundation, and Yale University Program in Reproductive Ecology.

References

Alden JC (1989) Osteoporosis—a review. Clin Ther 11(1):3–14

Aliabadi M, Kimiagar M, Ghayour-Mobarhan M, Shakeri MT, Nematy M, Ilaty AA, Moosavi AR, Lanham-New S (2008) Prevalence of malnutrition in free living elderly people in Iran: a cross-sectional study. Asia Pac J Clin Nutr 17(2):285–289

Balogh GA, Heulings R, Mailo DA, Russo PA, Sheriff F, Russo IH, Moral R, Russo J (2006) Genomic signature induced by pregnancy in the human breast. Int J Oncol 28(2):399–410

Bastian LA, West NA, Corcoran C, Munger RG (2005) Number of children and the risk of obesity in older women. Prev Med 40(1):99–104

Bererhi H, Kolhoff N, Constable A, Nielsen SP (1996) Multiparity and bone mass. Br J Obstet Gynaecol 103(8):818–821

Bernstein L (2002) Epidemiology of endocrine-related risk factors for breast cancer. J Mammary Gland Biol Neoplasia 7(1):3–15

Bertrand S, Alonso-Alvarez C, Devevey G, Faivre B, Prost J, Sorci G (2006) Carotenoids modulate the trade-off between egg production and resistance to oxidative stress in zebra finches. Oecologia 147(4):576–584

Biran A, Abbot J, Mace R (2004) Families and firewood: a comparative analysis of the costs and benefits of children in firewood collection and use in two rural communities in sub-Saharan Africa. Hum Ecol 32(1):1–25

Bisdee J, James W, Shaw M (1989) Changes in energy expenditure during the menstrual cycle. Br J Nutr 61:187–199

Black AJ, Topping J, Durham B, Farquharson RG, Fraser WD (2000) A detailed assessment of alterations in bone turnover, calcium homeostasis, and bone density in normal pregnancy. J Bone Miner Res 15(3):557–563

Blackburn S, Loper D (1992) Maternal, fetal, and neonatal physiology: a clinical perspective. W.B. Saunders Company, Philadelphia

Bray GA (2002) The underlying basis for obesity: relationship to cancer. J Nutr 132(11):3451S–3455S

Bronstein MN, Mak RP, King JC (1996) Unexpected relationship between fat mass and basal metabolic rate in pregnant women. Br J Nutr 75(5):659–668

Brown JE, Kaye SA, Folsom AR (1992) Parity-related weight change in women. Int J Obes (Lond) 16(9):627–631

Butte NF, King JC (2005) Energy requirements during pregnancy and lactation. Public Health Nutr 8(7A):1010–1027

Butte NF, Hopkinson JM, Mehta N, Moon JK, Smith EO (1999) Adjustments in energy expenditure and substrate utilization during late pregnancy and lactation. Am J Clin Nutr 69:299–307

Campbell C, Lee JZ (2004) Mortality and household in seven Liaodong populations, 1749–1909. In: Bengtsson T, Cameron C, Lee J (eds) Life under pressure. Mortality and living standards in Europe and Asia, 1700–1900. The MIT Press, Cambridge, MA, pp 293–324

Cole AH, Ibeziako PA, Bamgboye EA (1989) Basal metabolic rate and energy expenditure of pregnant Nigerian women. Br J Nutr 62(3):631–638

Cote K, Blanchard R, Lalumiere ML (2003) The influence of birth order on birth weight: does the sex of preceding siblings matter? J Biosoc Sci 35(3):455–462

Cummings SR, Nevitt MC, Browner WS, Stone K, Fox KM, Ensrud KE, Cauley JC, Black D, Vogt TM (1995) Risk factors for hip fracture in white women. N Engl J Med 332(12):767–773

Curtis V, Henry CJK, Birch E, Ghusain CA (1996) Intraindividual variation in the basal metabolic rate of women: effect of the menstrual cycle. Am J Hum Biol 8:631–639

De Aloysio D, Di Donato P, Giulini NA, Modena B, Cicchetti G, Comitini G, Gentile G, Cristiani P, Careccia A, Esposito E et al (2002) Risk of low bone density in women attending menopause clinics in Italy. Maturitas 42(2):105–111

Dribe M (2004) Long-term effects of childbearing on mortality: evidence from pre-industrial Sweden. Popul Stud J Demogr 58(3):297–310

Drinkwater BL, Chestnut CH (1991) Bone density changes during pregnancy and lactation in active women—a longitudinal study. Bone Miner 14(2):153–160

Durnin JVGA (1991) Energy requirements of pregnancy. Acta Paediatr Scand 373:33–42

Durnin JVGA (1993) Energy requirements in human pregnancy, in human nutrition and parasitic infection. Parasitology 107(l):S169–S175

Ellison PT, Lipson SF, O'Rourke MT, Bentley GR, Harrigan AM, Panter-Brick C, Vitzthum VJ (1993) Population variation in ovarian function. Lancet 342:433–434

Euler HA, Weitzel B (1996) Discriminative grandparental solicitude as reproductive strategy. Hum Nat 7(1):39–59

Filippini N, MacIntosh BJ, Hough MG, Goodwin GM, Frisoni GB, Smith SM, Matthews PM, Beckmann CF, Mackay CE (2009) Distinct patterns of brain activity in young carriers of the APOE-epsilon 4 allele. Proc Natl Acad Sci U S A 106(17):7209–7214

Forsum E, Kabir N, Sadurskis A, Westerterp K (1992) Total energy expenditure of healthy Swedish women during pregnancy and lactation. Am J Clin Nutr 56(2):334–342

Garner P, Smith T, Baea M, Lai D, Heywood P (1994) Maternal nutritional depletion in a rural area of Papua New Guinea. Trop Geogr Med 46(3):169–171

Goldberg GR, Prentice AM, Coward WA, Davies HL, Murgatroyd PR, Sawyer MB, Ashford J, Black AE (1991) Longitudinal assessment of the components of energy balance in well-nourished lactating women. Am J Clin Nutr 54(5):788–798

Green A, Beral V, Moser K (1988) Mortality in women in relation to their childbearing history. Br Med J 297(6645):391–395

Guillermo-Tuazon MA, Barba CV, van Raaij JM, Hautvast JG (1992) Energy intake, energy expenditure, and body composition of poor rural Philippine women throughout the first 6 mo of lactation. Am J Clin Nutr 56(5):874–880

Hamilton WD (1964) Genetical evolution of social behaviour, parts 1 and 2. J Theor Biol 7(1):1–52

Harris HE, Ellison GTH (1997) Do the changes in energy balance that occur during pregnancy predispose parous women to obesity? Nutr Res Rev 10:57–81

Hawkes K, O'Connell JF, Jones NGB (1997) Hadza women's time allocation, offspring provisioning, and the evolution of long postmenopausal life spans. Curr Anthropol 38(4):551–577

Hawkes K, O'Connell JF, Jones NGB, Alvarez H, Charnov EL (1998) Grandmothering, menopause, and the evolution of human life histories. Proc Natl Acad Sci U S A 95(3):1336–1339

Helle S, Lummaa V, Jokela J (2002) Sons reduced maternal longevity in preindustrial humans. Science 296:1085

Hinde K (2007) First-time macaque mothers bias milk composition in favor of sons. Curr Biol 17(22):R958–R959

Hinkula M, Pukkala E, Kyyronen P, Kauppila A (2001) Grand multiparity and the risk of breast cancer: population-based study in Finland. Cancer Causes Control 12(6):491–500

Hinkula M, Kauppila A, Nayha S, Pukkala E (2006) Cause-specific mortality of grand multiparous women in Finland. Am J Epidemiol 163(4):367–373

Hruschka DJ, Hagaman A (2015) The physiological cost of reproduction for rich and poor across 65 countries. Am J Hum Biol 27(5):654–659

Hytten FE, Leitch I (1971) The physiology of human pregnancy, 2nd edn. Blackwell, Oxford

Idler EL, Benyamini Y (1997) Self-rated health and mortality: a review of twenty-seven community studies. J Health Soc Behav 38(1):21–37

Jamison CS, Cornell LL, Jamison PL, Nakazato H (2002) Are all grandmothers equal? a review and a preliminary test of the "grandmother hypothesis" in Tokugawa Japan. Am J Phys Anthropol 119(1):67–76

Jasienska G (2001) Why energy expenditure causes reproductive suppression in women. An evolutionary and bioenergetic perspective. In: Ellison PT (ed) Reproductive ecology and human evolution. Aldine de Gruyter, New York, pp 59–85

Jasienska G (2003) Energy metabolism and the evolution of reproductive suppression in the human female. Acta Biotheor 51:1–18

Jasienska G (2013) The fragile wisdom. An evolutionary view on women's biology and health. Harvard University Press, Cambridge, MA

Jasienska G, Ellison PT (1998) Physical work causes suppression of ovarian function in women. Proc R Soc Lond B 265(1408):1847–1851

Jasienska G, Ellison PT (2004) Energetic factors and seasonal changes in ovarian function in women from rural Poland. Am J Hum Biol 16:563–580

Jasienska G, Thune I, Ellison PT (2000) Energetic factors, ovarian steroids and the risk of breast cancer. Eur J Cancer Prev 9:231–239

Jasienska G, Nenko I, Jasienski M (2006a) Daughters increase longevity of fathers, but daughters and sons equally reduce longevity of mothers. Am J Hum Biol 18(3):422–425

Jasienska G, Thune I, Ellison PT (2006b) Fatness at birth predicts adult susceptibility to ovarian suppression: an empirical test of the Predictive Adaptive Response hypothesis. Proc Natl Acad Sci U S A 103(34):12759–12762

Jasienska G, Ziomkiewicz A, Lipson SF, Thune I, Ellison PT (2006c) High ponderal index at birth predicts high estradiol levels in adult women. Am J Hum Biol 18(1):133–140

Jasienska G, Ellison PT, Galbarczyk A, Jasienski M, Kalemba-Drozdz M, Kapiszewska M, Nenko I, Thune I, Ziomkiewicz A (2015) Apolipoprotein E (ApoE) polymorphism is related to differences in potential fertility in women: a case of antagonistic pleiotropy? Proc R Soc B Biol Sci 282(1803)

Karlsson C, Obrant KJ, Karlsson M (2001) Pregnancy and lactation confer reversible bone loss in humans. Osteoporos Int 12(10):828–834

Karlsson MK, Ahlborg HG, Karlsson C (2005) Female reproductive history and the skeleton—a review. BJOG 112(7):851–856

Key TJA, Pike MC (1988) The role of oestrogens and progestagens in the epidemiology and prevention of breast cancer. Eur J Cancer 24:29–43

King JC, Butte NF, Bronstein MN, Koop LE, Lindquist SA (1994) Energy metabolism during pregnancy: influence of maternal energy status. Am J Clin Nutr 59:s439–s445

Kington R, Lillard L, Rogowski J (1997) Reproductive history, socioeconomic status, and self-reported health status of women aged 50 years or older. Am J Public Health 87(1):33–37

Kirchengast S (2000) Differential reproductive success and body size in !Kung San people from northern Namibia. Coll Antropol 24(1):121–132

Kramer KL (2002) Variation in juvenile dependence—helping behavior among Maya children. Hum Nat 13(2):299–325

Kramer KL (2005) Children's help and the pace of reproduction: cooperative breeding in humans. Evol Anthropol 14(6):224–237

Kvale G (1992) Reproductive factors in breast cancer epidemiology. Acta Oncol 31(2):187–194

Laskey MA, Prentice A (1997) Effect of pregnancy on recovery of lactational bone loss. Lancet 349(9064):1518–1519

Lawlor DA, Emberson JR, Ebrahim S, Whincup PH, Wannamethee SG, Walker M, Smith GD (2003) Is the association between parity and coronary heart disease due to biological effects of pregnancy or adverse lifestyle risk factors associated with child-rearing?: findings from the British women's heart and health study and the British regional heart study. Circulation 107(9):1260–1264

Lawrence M, Whitehead RG (1988) Physical activity and total energy expenditure in child-bearing Gambian women. Eur J Clin Nutr 42:145–160

Le Bourg E (2007) Does reproduction decrease longevity in human beings? Ageing Res Rev 6(2):141–149

Le Bourg E, Thon B, Legare J, Desjardins B, Charbonneau H (1993) Reproductive life of French-Canadians in the 17–18th-centuries — a search for a trade-off between early fecundity and longevity. Exp Gerontol 28(3):217–232

Leonard WR (2008) Lifestyle, diet, and disease: comparative perspectives on the determinants of chronic health risks. In: Stearns SC, Koella JC (eds) Evolution in health and disease. Oxford University Press, New York, pp 265–276

Little MA, Leslie PW, Campbell KL (1992) Energy reserves and parity of nomadic and settled Turkana women. Am J Hum Biol 4(6):729–738

Loos R, Derom C, Eeckels R, Derom R, Vlietinck R (2001) Length of gestation and birthweight in dizygotic twins. Lancet 358:560–561

Lund E, Arnesen E, Borgan JK (1990) Pattern of childbearing and mortality in married women — a national prospective study from Norway. J Epidemiol Community Health 44(3):237–240

Lunn PG (1994) Lactation and other metabolic loads affecting human reproduction. Ann N Y Acad Sci 709:77–85

Lycett JE, Dunbar RIM, Voland E (2000) Longevity and the costs of reproduction in a historical human population. Proc R Soc London B 267:31–35

Mace R, Sear R (1997) Birth interval and the sex of children in a traditional African population: an evolutionary analysis. J Biosoc Sci 29(4):499–507

MacMahon B (2006) Epidemiology and the causes of breast cancer. Int J Cancer 118(10):2373–2378

Madhavapeddi R, Rao BS (1992) Energy balance in lactating undernourished Indian women. Eur J Clin Nutr 46(5):349–354

Manor O, Eisenbach Z, Israeli A, Friedlander Y (2000) Mortality differentials among women: the Israel Longitudinal Mortality Study. Soc Sci Med 51(8):1175–1188

Marsal K, Persson P, Larsen T, Lilja H, Selbing A, Sultan B (1996) Intrauterine growth curves based on ultrasonically estimated foetal weights. Acta Paediatr 85:843–848

Martin JE, Hure AJ, Macdonald-Wicks L, Smith R, Collins CE (2014) Predictors of post-partum weight retention in a prospective longitudinal study. Matern Child Nutr 10(4):496–509

McArdle PF, Pollin TI, O'Connell JR, Sorkin JD, Agarwala R, Schaffer AA, Streeten EA, King TM, Shuldiner AR, Mitchell BD (2006) Does having children extend life span? a genealogical study of parity and longevity in the Amish. J Gerontol A Biol Sci Med Sci 61(2):190–195

McHaffie HE (1992) Social support in the neonatal intensive-care unit. J Adv Nurs 17(3):279–287

Meijer GAL, Westerterp KR, Saris WHM, Ten HF (1992) Sleeping metabolic rate in relation to body composition and the menstrual cycle. Am J Clin Nutr 55(3):637–640

Mettlin C (1999) Global breast cancer mortality statistics. CA Cancer J Clin 49(3):138–144

Michalski RL, Shackelford TK (2005) Grandparental investment as a function of relational uncertainty and emotional closeness with parents. Hum Nat 16(3):293–305

Muller HG, Chiou JM, Carey JR, Wang JL (2002) Fertility and life span: late children enhance female longevity. J Gerontol, Ser A 57:B202–B206

Nag M, White BNF, Peet RC (1978) Anthropological approach to study of economic value of children in Java and Nepal. Curr Anthropol 19(2):293–306

Ness RB, Harris T, Cobb J, Flegal KM, Kelsey JL, Balanger A, Stunkard AJ, Dagostino RB (1993) Number of pregnancies and the subsequent risk of cardiovascular disease. N Engl J Med 328(21):1528–1533

Nguyen TV, Jones G, Sambrook PN, White CP, Kelly PJ, Eisman JA (1995) Effects of estrogen exposure and reproductive factors on bone mineral density and osteoporotic fractures. J Clin Endocrinol Metab 80(9):2709–2714

Nunez-de la Mora A, Chatterton RT, Choudhury OA, Napolitano DA, Bentley GR (2007) Childhood conditions influence adult progesterone levels. PLoS Med 4(5), e167

Oria RB, Patrick PD, Oria MOB, Lorntz B, Thompson MR, Azevedo OGR, Lobo RNB, Pinkerton RF, Guerrant RL, Lima AAM (2010) ApoE polymorphisms and diarrheal outcomes in Brazilian shanty town children. Braz J Med Biol Res 43(3):249–256

Panter-Brick C, Ellison PT (1994) Seasonality of workloads and ovarian function in Nepali women. Ann N Y Acad Sci 709:234–235

Panter-Brick C, Lotstein DS, Ellison PT (1993) Seasonality of reproductive function and weight loss in rural Nepali women. Hum Reprod 8(5):684–690

Parikh NI, Cnattingius S, Dickman PW, Mittleman MA, Ludvigsson JF, Ingelsson E (2010) Parity and risk of later-life maternal cardiovascular disease. Am Heart J 159(2)

Pashos A (2000) Does paternal uncertainty explain discriminative grandparental solicitude? a cross-cultural study in Greece and Germany. Evol Hum Behav 21(2):97–109

Peacock N (1991) An evolutionary perspective on the patterning of maternal investment in pregnancy. Hum Nat 2:351–385

Piers LS, Diggavi SN, Thangam S, Van RJMA, Shetty PS, Hautvast JGAJ (1995) Changes in energy expenditure, anthropometry, and energy intake during the course of pregnancy and lactation in well-nourished Indian women. Am J Clin Nutr 61(3):501–513

Prentice AM (2000) Calcium in pregnancy and lactation. Annu Rev Nutr 20:249–272

Prentice AM, Prentice A (1990) Maternal energy requirements to support lactation. In: Atkinson SA, Hanson LA, Chandra RK (eds) Breastfeeding, nutrition, infection and infant growth in developed and emerging countries. ARTS Biomedical Publishers and Distributors, St. John's, Nfld, pp 69–86

Prentice AM, Whitehead RG (1987) The energetics of human reproduction. Symp Zool Soc London 57:275–304

Prentice AM, Goldberg GR, Davies HL, Murgatroyd PR, Scott W (1989) Energy-sparing adaptations in human pregnancy assessed by whole-body calorimetry. Br J Nutr 62(1):5–22

Prentice AM, Poppitt SD, Goldberg GR, Prentice A (1995) Adaptive strategies regulating energy balance in human pregnancy. Hum Reprod Update 1(2):149–161

Puskarczyk K, Galbarczyk A, Klimek M, Nenko I, Odrzywolek L, Jasienska G (2015) Being born after your brother is not a disadvantage: reproductive success does not depend on the sex of the preceding sibling. Am J Hum Biol 27(5):731–733

Qureshi AI, Giles WH, Croft JB, Stern BJ (1997) Number of pregnancies and risk for stroke and stroke subtypes. Arch Neurol 54(2):203–206

Raisz LG (2005) Pathogenesis of osteoporosis: concepts, conflicts, and prospects. J Clin Invest 115(12):3318–3325

Rashid M, Ulijaszek SJ (1999) Daily energy expenditure across the course of lactation among urban Bangladeshi women. Am J Phys Anthropol 110(4):457–465

Rickard IJ (2008) Offspring are lighter at birth and smaller in adulthood when born after a brother versus a sister in humans. Evol Hum Behav 29(3):196–200

Rickard IJ, Russell AF, Lummaa V (2007) Producing sons reduces lifetime reproductive success of subsequent offspring in pre-industrial Finns. Proc R Soc B Biol Sci 274(1628):2981–2988

Sear R, Mace R (2008) Who keeps children alive? a review of the effects of kin on child survival. Evol Hum Behav 29(1):1–18

Sear R, Mace R, McGregor IA (2000) Maternal grandmothers improve nutritional status and survival of children in rural Gambia. Proc R Soc Lond B 267(1453):1641–1647

Simmons D, Shaw J, McKenzie A, Eaton S, Cameron AJ, Zimmet P (2006) Is grand multiparity associated with an increased risk of dysglycaemia? Diabetologia 49(7):1522–1527

Somner J, McLellan S, Cheung J, Mak YT, Frost ML, Knapp KM, Wierzbicki AS, Wheeler M, Fogelman I, Ralston SH et al (2004) Polymorphisms in the p450 c17 (17-hydroxylase/17,20-lyase) and p450 c19 (aromatase) genes: association with serum sex steroid concentrations and bone mineral density in postmenopausal women. J Clin Endocrinol Metabol 89(1):344–351

Song YQ, Stampfer MJ, Liu SM (2004) Meta-analysis: apolipoprotein E genotypes and risk for coronary heart disease. Ann Intern Med 141(2):137–147

Stearns SC (1989) Trade-offs in life-history evolution. Funct Ecol 3(3):259–268

Stieglitz J, Beheim BA, Trumble BC, Madimenos FC, Kaplan H, Gurven M (2015) Low mineral density of a weight-bearing bone among adult women in a high fertility population. Am J Phys Anthropol 156(4):637–648

Strassmann BI (1996a) Energy economy in the evolution of menstruation. Evol Anthropol 5:157–164

Strassmann BI (1996b) The evolution of endometrial cycles and menstruation. Q Rev Biol 71:181–220

Streeten EA, Ryan KA, McBride DJ, Pollin TI, Shuldiner AR, Mitchell BD (2005) The relationship between parity and bone mineral density in women characterized by a homogeneous lifestyle and high parity. J Clin Endocrinol Metabol 90(8):4536–4541

Sujatha T, Shatrugna V, Venkataramana Y, Begum N (2000) Energy expenditure on household, childcare and occupational activities of women from urban poor households. Br J Nutr 83(5):497–503

Tracer DP (1991) Fertility-related changes in maternal body composition among the au of Papua New Guinea. Am J Phys Anthropol 85(4):393–406

Tsuya NO, Kurosu S, Nakazato H (2004) Mortality and household in two Ou villages 1716–1870. In: Bengtsson T, Cameron C, Lee JZ (eds) Life under pressure. Mortality and living standards in Europe and Asia, 1700–1900. The MIT Press, Cambridge, MA, pp 253–292

Valeggia CR, Ellison PT (2003) Impact of breastfeeding on anthropometric changes in peri-urban Toba women (Argentina). Am J Hum Biol 15(5):717–724

Valeggia CR, Ellison PT (2004) Lactational amenorrhoea in well-nourished Toba women of Formosa, Argentina. J Biosoc Sci 36(5):573–595

Van de Putte B, Matthijs K, Vlietinck R (2003) A social component in the negative effect of sons on maternal longevity in pre-industrial humans. J Biosoc Sci 36:289–297

Vandenheede H, Deboosere P, Gadeyne S, De Spiegelaere M (2012) The associations between nationality, fertility history and diabetes-related mortality: a retrospective cohort study in the Brussels-Capital Region (2001–2005). J Public Health 34(1):100–107

Voland E, Beise J (2002) Opposite effects of maternal and paternal grandmothers on infant survival in historical Krummhorn. Behav Ecol Sociobiol 52(6):435–443

Winkvist A, Rasmussen KM, Habicht JP (1992) A new definition of maternal depletion syndrome. Am J Public Health 82(5):691–694

Zera AJ, Harshman LG (2001) The physiology of life history trade-offs in animals. Annu Rev Ecol Syst 32:95–126

Chapter 11
From Novel to Extreme: Contemporary Environments and Physiologic Dysfunction

Diana S. Sherry

Abstract This chapter presents a new conceptual framework to characterize the interactions between modern environments and ancestral physiology that can influence health over the life course. Although the human body was not designed by natural selection to maximize health, it was nonetheless designed to function within certain environmental parameters. Physiology serves as the essential interface between genes and environment not only during life history transitions but also during short-term responses needed to regulate the impact of environmental variation on the internal state. Here, I argue that the contemporary habitat has become an "extreme environment" in the sense that, much like climbing Mount Everest, it requires the human body (and psyche) to function beyond the limits of its adaptive capacity, with potential dire consequences to health. Adaptive mechanisms that have become dysfunctional in an extreme environment show three distinct features: (1) gradient effects, often without overt signs of dysfunction, whereby (2) compensatory mechanisms themselves become the source of illness, and (3) involve systemic repercussions. I present an analysis of altitude hypoxia to illustrate an extreme environmental condition and then apply the same framework to a consideration of metabolic disorders. A conceptual framework that identifies lifestyle factors as being equivalent to entering an extreme environment conveys an immediate, intuitive sense of urgency and an implicit recognition that the human organism has strayed into a habitat for which it is ill equipped.

11.1 Introduction

Evolutionary perspectives on health and medicine have given rise to two prominent conceptual models and messages (Stearns et al. 2010). The first emphasizes inherent design "flaws" in the human body arising from inevitable life history trade-offs

D.S. Sherry (✉)
School of Communication, Institute of Liberal Arts and Interdisciplinary Studies, Emerson College, 120 Boylston St., Boston, MA 02116, USA

Department of Human Evolutionary Biology, Harvard University, 11 Divinity Ave, Cambridge, MA 02138, USA
e-mail: Diana_Sherry@emerson.edu

involved in maximizing lifetime reproductive success. The second examines physiologic variation along dimensions of health in terms of a "functional continuum" composed of adaptive responses to environmental conditions. Both approaches recognize the growing evidence of widespread health risks associated with the dysregulation of major biological systems—metabolic, cardiovascular, and reproductive—under the influence of modern environments.

Typically, the modern environment has been characterized as a novel one, generating a "mismatch" between ancestral genes and contemporary Western lifestyles (Nesse and Williams 1994; Lieberman 2003; Pollard 2008; Trevathan et al. 2008). Although the term mismatch has been referred to as a hypothesis (Lieberman 2013) and a paradigm (Gluckman and Hanson 2006), it is clear that the term serves to describe the evolutionary basis for many of the contemporary trends in disease (Ewald 1994; Panter-Brick and Worthman 1999; Greaves 2000; Stearns and Koella 2008). Yet, as a heuristic and communication device, mismatch falls short for two reasons. First, the concept of mismatch connotes a relationship between two discrete entities (Willingham 2009), neglecting to capture or emphasize the important gradient effects that arise from the interaction between ancestral genes and modern environments. Second, as an emotionally neutral term, mismatch fails to convey the seriousness of health risks associated with the unraveling of biological systems (dysregulation) under contemporary conditions, especially those involving lifestyle factors.

The dysregulation of physiologic systems resulting from gene–environment interaction tends to fall along a graded continuum, whereby clinically recognized illness and pathology manifest after a period of relatively modest forms of dysregulation. This type of dysfunction resembles what happens to the human body when it enters an extreme environment. Although humans have occupied a broad range of habitats over the course of evolution, I argue that the contemporary habitat has become an "extreme environment" in the sense that, much like climbing Mount Everest, it requires the human body to function beyond the limits of its adaptive capacity, with potential dire consequences to health.

As Bateson (1987, p. 349) noted, a "lethal change in either environment or genotype is simply one which demands somatic modifications which the organism cannot achieve." Under extreme conditions, not only do somatic modifications reach their limit in terms of effectiveness, but the compensatory mechanisms themselves begin to wreak havoc, although the body is performing exactly as designed by natural selection. Conversely, evidence of compensatory mechanisms as the source of illness signifies that the organism is attempting to operate under extreme conditions. Adaptive mechanisms that have become dysfunctional in an extreme environment show three distinct features: (1) gradient effects, often without overt signs of dysfunction, whereby (2) compensatory mechanisms themselves become the genesis of illness, and (3) involve systemic repercussions.

Bateson (1987) also pointed out in his essay first published in 1963 that somatic flexibility is built into the genotype as long as the benefits of flexibility outweigh the fitness costs. Evidence of a flexible functional continuum, such as the ovarian function response continuum (Ellison 1990; Ellison 2001), shifts the perception of pathology in relation to adaptive physiological mechanisms. Here, I propose that a "dysfunctional continuum" emerges whenever a departure from ancestral environmental conditions requires the human body (and psyche) to function in the equivalent of an extreme environment. I first consider the context of altitude hypoxia as an extreme environmental parameter and synthesize the research presented in West et al. (2013) to assemble the dysfunctional continuum en route to clinically recognized pathology. I then apply the same framework to a consideration of metabolic disorders, using standard textbook physiology to describe the foundational narratives of metabolism (Neal 2001; Dominiczak 2007).

From an evolutionary perspective, the current clinical view of functional physiology can mask potential dysfunction arising from adaptive responses to extreme conditions and thus requires a shift in perception. For example, altitude acclimatization is not a benign process and, for the most part, does not restore physiological systems to the sea level equivalent of functionality. Instead, acclimatization capitalizes on somatic flexibility by co-opting adaptive mechanisms designed to function at sea level (and roughly below 2000–2500 m). Although the body may not show overt signs of distress, systems are nonetheless strained and unraveling and, depending on the duration and magnitude of hypoxia, will continue to deteriorate along a continuum of dysfunction. Rather than a series of cost-free adjustments with little consequences to health, acclimatization represents the first line of defense against environmental assault.

11.2   Hypoxia

11.2.1   Adaptive Physiology at Sea Level

Two types of peripheral sensors monitor oxygen availability in the blood. The carotid body, located above the bifurcation in each carotid artery, acts as an immediate transducer if the arterial oxygen pressure drops, converting the hypoxic signal from the blood to a neural signal to the brain. Within seconds, the brain sends impulses to the respiratory muscles to increase the rate and depth of breathing (ventilation). Deep and rapid breathing opens up dormant alveoli and increases both the amount of air entering the lungs and the rate of oxygen diffusion across the capillary membranes. The hypoxic ventilatory response (HVR) is accompanied by a number of complementary responses, including an increase in heart rate and pulmonary artery pressure (caused by greater contraction and force of the right heart ventricle) and also an increase in cerebral blood flow to maintain a steady state of oxygen delivery to the brain.

Another rapid response stimulated by the carotid–neural pathway is a decrease in plasma volume orchestrated by the kidney (via shifts in sodium reabsorption), quickly boosting the supply of oxygen carried per unit volume of blood by increasing the concentration of red blood cells and hemoglobin. In addition, the enzyme that degrades the transcriptional factor HIF-1α (continuously formed and degraded in tissues throughout the body) is inhibited under conditions of hypoxia within minutes, facilitating the induction of certain genes and, in particular, the upregulation of vascular endothelial growth factor (VEGF), known to play an important role in wound healing among other vascular tasks.

Oxygen-sensing chemoreceptors also exist in the kidney but serve a different purpose than the carotid bodies. Designed primarily to respond to hypoxic signals of injury and blood loss, the kidney monitors oxygen content (the quantity attached to hemoglobin) and stimulates the production of red blood cells in the bone marrow by releasing the hormone erythropoietin (EPO). Although EPO levels rise in response to hypoxic conditions within hours, the full process of increasing red blood cell mass takes a period of weeks.

The lungs are also equipped with a means to defend against hypoxia resulting from injury or disease at sea level. Pulmonary vasoconstriction in response to local tissue hypoxia closes off the damaged section of the lung through the contraction of the smooth muscle in the small pulmonary arteries and effectively redirects blood flow to the undamaged regions of the lung.

While the peripheral sensors monitor the status of oxygen in the blood, a central mechanism located in the respiratory center of the brainstem uses the status of carbon dioxide (sensed primarily by changes in blood pH) to help regulate the processes of ventilation and cerebral blood flow. At sea level, the peripheral and central chemoreceptors provide complementary signals on the status of respiration. An increase in physical exertion, for example, typically requires both an increase in the supply of oxygen and a high rate of disposal of carbon dioxide, a by-product of the oxidative activity in tissues. An increase in carbon dioxide thus indicates that respiration rate needs to be ratcheted up along with cerebral vasodilation to meet the demands of increased physical activity or danger.

11.2.2 Acute Mountain Sickness

Altitude poses a potential threat to the human organism because the concentration of oxygen (roughly 21 %) in the atmosphere remains constant, although the barometric pressure decreases as altitude increases. The drop in barometric pressure translates into less oxygen per breath, referred to as a drop in the *partial* pressure of oxygen (PO_2) because each gas exerts pressure based on its concentration independent of other gases present.

Acute exposure to moderate altitudes (above 2000–2500 m), such as flying from sea level to the Rocky Mountains for a ski vacation in Colorado, can cause acute mountain sickness (AMS). The drop in PO_2 registers with the peripheral chemoreceptors within a few minutes after ascent, initiating the hypoxic ventilatory response (HVR). Yet, the onset of AMS can be delayed for another 6–24 h, indicating that when the HVR does not ameliorate the hypoxia, other responses give rise to the symptoms of illness. Common symptoms include headache, fatigue, gastrointestinal ailments (loss of appetite, nausea, vomiting), and a state of expanded extracellular fluid (evidence that the kidneys are responding by reducing plasma volume). AMS symptoms most likely reflect mild generalized brain swelling commensurate with increases in cerebral blood flow in an effort to oxygenate the brain. A swelling brain pressing against a rigid braincase likely leads to the headache and other AMS symptoms, although the precise physiologic pathways have not been fully determined. AMS does not pose a serious threat but can be prevented or mitigated by a gradual ascent, which allows time for other compensatory mechanisms to respond and maintain oxygen delivery to the brain.

11.2.3 Acclimatization as Deterioration

Ascending to higher elevations becomes possible through additional mechanisms designed to respond to hypoxic conditions at sea level. The series of physiologic changes that mitigate the drop in PO_2 constitutes the process of acclimatization, a term that refers only to changes "seen as beneficial as opposed to changes which result in illness such as AMS (West et al. 2013, p. 57)." From an evolutionary perspective, altitude acclimatization falls outside the functional continuum and represents a state of physiologic dysfunction, progressing along a gradient of deterioration in response to the magnitude and duration of exposure to an extreme condition.

Further exposure to altitude continues to call upon the body's repertoire of responses to signals of hypoxia as if operating at sea level. Stimulation of red blood cell production under the influence of EPO (erythropoietin) continues, increasing the hemoglobin concentration [Hb] per unit volume of blood. Plasma volume usually returns to sea level values after a few weeks. The increase in [Hb] tends to rise linearly with altitude (to about 5300 m) and generally restores the oxygen carrying content of the blood to near sea level values and also generates adjustments in blood flow so that oxygen delivery to the brain, in particular, remains constant over a range of [Hb]. At the same time, the rise in [Hb] thickens the blood, increasing vascular resistance throughout the circulation and contributing to the already raised pulmonary artery pressure. In addition, the rise in [Hb] does not translate into improvements in the work capacity of muscles, as it does at sea level, because oxygen diffusion at the tissue site rather than oxygen delivery becomes more limited at altitude.

Structural changes also begin to occur in the pulmonary vasculature in response to the rise in pulmonary artery pressure sensed in the vascular endothelium (lining of blood vessels). Referred to as remodeling, bands of smooth muscle develop fairly rapidly in the small pulmonary arteries under the influence of VEGF. Although vascular remodeling reduces stress on blood vessel walls, resistance is exacerbated by the narrowing of the vessel tube or lumen. Unlike vasoconstriction and the simple contraction of the vascular smooth muscle, pulmonary remodeling introduces structural changes with a substantial degree of irreversibility.

11.2.4 Corrupted Signals

As the human organism ascends along the altitude gradient, physiologic systems continue to be strained in an effort to respond to internal signals of distress that have lost their functional integrity. Key signals become corrupted in the sense that they no longer convey accurate information, as a consequence of both the shifting external parameters and the internal efforts to remedy the situation. In other words, in an extreme environment, the functional responses themselves become dysfunctional and also become the source of further damage that may or may not be reversible, depending on the degree and duration of exposure.

Two potentially fatal forms of AMS—high-altitude pulmonary edema (HAPE) and high-altitude cerebral edema (HACE)—provide dramatic illustrations of the phenomenon of corrupted signals. HAPE, characterized by frothy sputum that may become blood tinged, can develop at relatively moderate altitude (as low as 2000 m). The genesis of HAPE lies in the main compensatory response by the lungs to the alarm signal of hypoxia. The pulmonary vasoconstriction mechanism, effective at sea level, becomes invoked more widely throughout the lungs yet unevenly at altitude, forcing total cardiac output through fewer sections of the lung as pulmonary arteries progressively close off. Given the high pulmonary artery pressure (PAP) at altitude, capillary walls can break or burst, leaking blood fluids into the alveolar space. A person literally begins to drown. Capillary walls are the most vulnerable structures in the pulmonary vasculature and must be thin enough to allow for gas exchange yet strong enough to withstand increases in pulmonary artery pressure during physical exertion. (Bleeding into the lungs has been found in racehorses in relation to very high PAP during physical exertion.) Although physical exertion at altitude can quickly increase the risk of developing HAPE, breaks in the capillary walls are readily repaired, and most individuals improve rapidly upon descent to lower altitude.

High-altitude cerebral edema (HACE) also arises from breaks in the capillary walls under pressure but tends to occur at higher altitude (above 3500 m). Symptoms include loss of balance, hallucinations, and hostile or irrational behavior. HACE symptoms stem from the compensatory drive to increase blood flow to the brain in response to the unremitting signals of hypoxia, generating an inordinate amount of swelling and pressure, compounded further by damage to the capillary walls and the subsequent leak of blood fluids into the extravascular space. Descent to lower altitude becomes imperative as death can occur within a matter of days or hours.

Less dramatically, signals to the brain on the status of respiration also become corrupted at altitude. The prolonged HVR leads to a disproportional loss of carbon dioxide (CO_2) because carbon dioxide diffuses through tissues about 20 times faster than oxygen due to its higher solubility. The different rates of diffusion for the two gases mean that CO_2 is being eliminated through expiration faster than it is being produced by tissues, causing a shift in blood pH toward greater alkalinity. At sea level, alkaline blood pH signals the brain to lower the rate of ventilation and reduce cerebral blood flow through vasoconstriction. At altitude, the brain receives conflicting signals because low arterial PO_2 simultaneously signals the brain to increase ventilation and cerebral vasodilation. Low PCO_2 no longer conveys reliable complementary information at altitude, creating an unsteady state not encountered at sea level. Although the hypoxic drive to increase ventilation overrides the low PCO_2 signal, the mixed signals to the brain may introduce some difficulty regulating cerebral blood flow. Irregular cerebral vasoconstriction, leading to diminished oxygen delivery to the brain, may contribute to the impairment of cognitive functions seen at altitude, including reaction time and hand–eye coordination as well as memory and speech.

Fluctuations in blood pH provide an index of PCO_2 level because CO_2 travels in the bloodstream attached to bicarbonate (an alkaloid) until expelled by the lungs. A drop in PCO_2 leads to a rise in unattached bicarbonate, requiring a compensatory adjustment to ensure that body fluids maintain nearly neutral pH. The kidneys manage the correction fairly well at altitude by increasing the rate of bicarbonate excretion. However, at very high altitude (above 6500 m), the renal compensatory mechanism no longer functions effectively because the frequency of urination needed to maintain neutral pH would require the depletion of body fluids and lead to dehydration. Instead, the kidneys slow down the rate of bicarbonate excretion to preserve body fluids, and blood pH becomes severely alkaline—a disorder called respiratory alkalosis.

Skeletal muscles also initiate a suite of compensatory mechanisms at altitude to improve oxidative ability, as if undergoing greater demands for physical activity similar to endurance training over the course of several weeks at sea level. Under the stimulus of local tissue hypoxia, some responses improve oxygen delivery. For example, the proliferation of new capillaries facilitated by VEGF increases the surface area for oxygen diffusion and simultaneously decreases the diffusion distance to mitochondria, the primary sites of oxygen utilization. Oxidative capacity also improves due to increases in the total tissue volume of mitochondria and the augmented activity of key enzymes of the Krebs cycle and electron transport chain.

In spite of these local tissue changes, low tolerance for muscle activity, a prominent feature of exposure to altitude, persists. One contributing factor is that the muscles of locomotion receive less blood flow in deference to the muscles of respiration (primarily the diaphragm); the HVR requires a greater proportion of cardiac output for the activity of breathing than at sea level, thereby limiting the perfusion of blood to skeletal muscles and reducing the effectiveness of local tissue changes. Another contributing factor is that the anaerobic mechanism for supplying fuel and supporting muscle activity at sea level is not invoked at altitude. Known as the "lactate paradox" (in reference to the anaerobic by-product of lactic acid), the anaerobic pathway remains intact but not utilized, apparently held in check by the local tissue signals of enhanced oxidative ability. Notably, at relatively high altitude (above 5300–6000 m), local tissue changes wane, and muscle becomes preferentially lost over fat, in contrast to weight loss in the form of fat at moderate altitude. A reduction in the size and diameter of muscle fibers at high altitude increases capillary density (the number of capillaries per unit volume of tissue) as a way to improve oxygen diffusion at the tissue site.

11.2.5   Chronic Illness

For populations living at or commuting routinely to altitude above 2500 m, with the notable exception of Tibetans and some Andean populations, chronic exposure to hypoxic conditions over a period of months to years can manifest clinically as two types of illness: chronic mountain sickness (CMS) and high-altitude pulmonary hypertension (HAPH). CMS tends to occur in the high-altitude areas of South America and also has been found in Leadville, Colorado (3100 m), whereas HAPH tends to occur in the high-altitude regions of Asia. Although symptoms for both conditions show considerable overlap, CMS denotes an excessive number of red blood cells, while HAPH refers to the predominance of pulmonary hypertension.

In the early phase of CMS, red blood cell mass can increase by as much as 50 % of sea level values, although the benefits remain limited and both mental and physical deterioration ensue over time. Main symptoms include headache, loss of mental acuity, fatigue, and muscle weakness. The later stages of the illness involve hypertrophy of the right heart ventricle and moderate to severe pulmonary hypertension. Pulmonary hypertension develops not only because of the rise in blood viscosity but also due to the lung's two compensatory mechanisms: arterial vasoconstriction and pulmonary remodeling. Ultimately, the compensatory mechanisms operating in both CMS and HAPH throw great strain on the right heart and can lead to congestive heart failure. Unfortunately, mild to moderate pulmonary hypertension does not give rise to symptoms. When signs do appear, including headache, labored breathing, sleeplessness, and irritability, the right heart has already begun to fail.

Tibetans, in particular, given their long history of residing at high altitude, appear to have undergone a degree of physiologic adaptation and tend to show very little rise in pulmonary artery pressure (similar to altitude-adapted animals such as the yak) under hypoxic conditions. The lack of pulmonary hypertension is likely related, in part, to the prominent upregulation of nitric oxide production compared to other populations (see Beall et al. 2012). Nitric oxide, a powerful vasodilator, presumably facilitates sufficient oxygen delivery to tissues.

Although constitutional factors, such as age and genetic polymorphisms, contribute to the likelihood of developing CMS and HAPH, chronic exposure to physiologic mechanisms designed as short-term remedies for hypoxia at sea level, and with limited effectiveness at altitude, generates dysfunction along a graded continuum that manifests as poor health only once the condition has become severe and potentially life-threatening.

11.3   Metabolic Dysfunction

The metabolic system was designed by natural selection to accomplish three main tasks: convert food into energy, convert surplus energy into storage, and mobilize energy from storage when needed. Because the body requires a continuous supply of energy, especially in the form of glucose (the only fuel the brain normally uses), the existence of energy storage pathways allows the body to function without a constant intake of food. Surplus glucose can be repackaged and stored in the liver and skeletal muscle as glycogen and as triglycerides in fat tissue. Managing the metabolic cycles of feeding and fasting falls primarily under the purview of the pancreas, orchestrated via changes in the production of the hormones insulin and glucagon by specialized beta cells and alpha cells, respectively. When food supply ceases, energy metabolism shifts direction within a few hours, reconfigured by a change in the ratio of insulin to glucagon circulating in the bloodstream. The adrenal glands also play a role in the signaling system.

11.3.1   Glycogen Stasis

The human body was also designed for relatively prolonged periods of physical activity followed by rest (e.g., Bramble and Lieberman 2004). The lack of physical activity, especially involving the use of major muscles such as the quadriceps, sets the stage for subsequent metabolic unraveling along a dysfunctional continuum. Muscle contractions depend upon the oxidation of the two major metabolic fuels, glucose and fatty acids, to generate ATP for energy. Initial physical activity draws upon the available ATP pool, but within minutes skeletal muscle begins to rely on glucose derived from its glycogen stores. Although glycogen can support muscle activity for about an hour, fatty acids become the main fuel for prolonged physical activity, mobilized after about 15–20 min by the low insulin-to-glucagon ratio. Skeletal muscle at rest also relies on fatty acid oxidation.

When major muscles are not being used for sustained activity, glycogen stores no longer need to be replenished to the degree required by a body experiencing states of high energy utilization or flux. When the cycles of utilization and replacement of glycogen, in particular, dampen or cease altogether, the physiologic mechanisms for metabolic regulation begin to operate in ways that depart from the ancestral state. Under conditions of chronic inactivity, "glycogen stasis" ensues, prompting muscles to adjust accordingly. Recognized clinically as a state of insulin resistance, skeletal muscles downregulate their responsiveness to insulin in order to match the low levels of physical activity beyond baseline maintenance. Downregulation primarily involves reducing the number of insulin receptors on the cell surface, thereby interfering with insulin's binding capacity and ability to command glucose uptake by the cell. Binding sets in motion a series of signaling cascades within the cell that recruit and direct the glucose transporter (GLUT) to move to the surface and usher glucose into the cell. Tissue-specific glucose transporters include GLUT4 for skeletal muscle and adipose tissue, GLUT5 for the intestines, and GLUT2 for the kidneys and pancreatic beta cells.

11.3.2   Glucose Overload

In addition to glycogen stasis, a second major metabolic departure from the formative ancestral past is the consumption of high glycemic foods and drinks on a consistent basis. After a meal containing carbohydrates, most (80%) of the circulating glucose is absorbed by tissues that do not require insulin for glucose uptake, mainly the brain (and erythrocytes). Uptake of the remaining peripheral glucose occurs mostly in insulin-dependent tissues, primarily skeletal muscle (85%) and adipose tissue (10%).

Not surprisingly, given the strong selective pressure to safeguard a steady supply of glucose to the brain, the human body comes well equipped with an array of physiologic mechanisms designed to compensate for stints of mild hypoglycemia. Mild hypoglycemia was probably encountered regularly in the ancestral past, not only on a short-term scale of hours, such as between feeding bouts and during physical activity, but also on a longer-term scale in the case of injury or infection extending to days and weeks.

The two principle regulators of hypoglycemia are glucagon and epinephrine (also called adrenaline, produced by the adrenal glands). Both hormones activate the transformation of glycogen stores into available glucose, with glucagon acting specifically on the liver and epinephrine acting on muscle tissue. Glucose derived from liver glycogen can be released readily into the bloodstream to fuel the brain and other tissues. Notably, glucose derived from muscle glycogen remains designated for muscle use only; muscles do not possess the enzyme (glucose-6-phosphatase) required to release glucose from cells. Along with activating the degradation of glycogen, glucagon and epinephrine stimulate the production of lipase, a key enzyme that promotes lipolysis and the breakdown of stored triglycerides into free fatty acids and glycerol. Although fatty acids cannot be converted into glucose, the liver can transform glycerol into glucose for release into the general circulation. In addition, glucagon instructs the liver to produce more low density lipoproteins (LDLs) needed to transport free fatty acids to muscle tissue, thereby ensuring that muscles receive the requisite fuel for sustained activity.

In contrast, insulin is the only hormone responsible for removing glucose from the bloodstream. This fact indicates that the difficulty of eliminating excess glucose was probably not encountered sufficiently in the ancestral past to warrant selection for multiple regulators. Physiologic asymmetries such as this are not unusual and tend to reflect the relative strength of past selective pressures. The human body, for example, possesses multiple mechanisms to avoid overheating and relatively few in response to cold.

The consumption of high glycemic foods and drinks typically invokes a robust but imprecise insulin response, referred to as "overshooting" on the part of the pancreas. (Mild hypoglycemia often follows, experienced as being hungry again.) This imprecise insulin response to high glycemic loads suggests a domain of somatic flexibility: selective pressures were not strong enough to require a precise calibration. In this way, the human organism could take full advantage of sporadic, short-lived (acute) opportunities to consume high glycemic foods, such as during honey season, without altering the existing mechanism—as long as the costs of imprecision remained relatively low. However, physiologic mechanisms not highly canalized by natural selection are also highly vulnerable to dysfunction.

Where does a deluge of glucose go? When glycogen stores are full, surplus glucose will be directed to fat tissue for storage. Since skeletal muscle is the primary target tissue for most of the peripheral glucose disposal after a meal, glycogen stasis combined with chronic exposure to repeated bouts of glucose overload inevitably leads to steady weight gain. In a study based on imaging technology, Petersen et al. (2007) demonstrated that lean subjects with insulin resistance in skeletal muscle tended to convert excess glucose from high carbohydrate meals directly to triglycerides rather than glycogen, unlike the lean normal control group. Although periodic or seasonal fluctuations in body fat characterize the nutritional ecology of foragers (e.g., Sherry and Marlowe 2007), steady or chronic increases in fat stores represent an overt sign that the body has stepped onto the path of metabolic dysfunction.

When viewed as a symptom of metabolic dysfunction already underway rather than a cause, steady weight gain only compounds the complications. For example, as an active endocrine organ, adipose tissue secretes various products (adipokines), including tumor necrosis factor-α (TNF-α). Notably, TNF-α acts to decrease the expression of GLUT4, the glucose transporter in muscle tissue. If seen as an adaptive mechanism, this pathway could allow the body to build or restore fat reserves during periods of energy abundance (Pond 1998; Wells 2010) by directing a greater proportion of peripheral glucose uptake to adipose tissue. Yet, this pathway also helps to explain how chronic weight gain can exacerbate insulin resistance in muscle tissue, thereby contributing to the further accumulation of fat stores. If left to continue unabated, this metabolic state of affairs sets in motion a number of far-reaching deleterious effects.

11.3.3   Compensatory Hazards

How does the body deal with chronic exposure to repeated bouts of glucose overload in the face of glycogen stasis? As the downregulation of glucose uptake by muscles begins to slow down the rate of clearance, excess glucose remains in the circulation for longer periods, triggering the pancreatic beta cells to ratchet up the production and dispatch of greater quantities of insulin—but to little avail. A cruel irony follows: circulating levels of glucose continue to creep higher and higher in the presence of ever-increasing outputs of insulin.

Over time, the ongoing frustrated effort to compensate for creeping hyperglycemia itself becomes the source of dysfunction, as the pancreas begins to fail. Although the precise pathways are not fully understood, beta cells lose their capacity to manufacture insulin, attributed to beta cell exhaustion or "burnout" and a sensitivity to oxidative damage (Hennige et al. 2003; Rashidi et al. 2009). (The phenomenon of organ exhaustion is not unique to the pancreas and can occur in the adrenal glands.) With the onset of incapacitated beta cells, the insulin-to-glucagon ratio becomes a corrupted signal, eliciting compensatory responses to a perceived shortage of glucose that doesn't exist. Plenty of glucose remains in the bloodstream, yet a sham state of hypoglycemia signaled by the predominance of glucagon stimulates the hepatic release of glucose from glycogen stores, thereby contributing to the escalating hyperglycemia. Not surprisingly, LDLs and fatty acids also pour into the bloodstream under the influence of glucagon, further compounding the repercussions.

11.3.4   Clinical Manifestations and Systemic Effects

The unraveling of metabolic regulation described above along a dysfunctional continuum maps onto the typology of clinically recognized metabolic disorders as follows: The initial appearance of high levels of glucose due to slow clearance (insulin resistance) is classified as impaired glucose tolerance (IGT). Extremely high levels of circulating glucose (hyperglycemia) coupled with high compensatory insulin levels (hyperinsulinemia) define the Metabolic Syndrome. Beta cell failure, the metabolic equivalent of entering "the death zone" on Everest (above 8000 m), becomes recognized as Type-2 Diabetes (T2D), usually accompanied by disordered lipid metabolism (hyperlipidemia, one of the prevailing effects of the compensatory hypoglycemic response) and the risk of long-term vascular complications. The relationship between metabolic disorders and cardiovascular disease takes hold in the systemic effects of metabolic dysfunction well before any clinical manifestations.

11.3.4.1   Cardiovascular Health

Metabolism comes at a cost even under the best of circumstances. The oxidation or burning of metabolic fuels (glucose and fatty acids) to release energy as ATP yields a number of deleterious by-products classified into two main types: reactive oxygen species (ROS) known to damage DNA and advanced glycation end products (AGEs) known to damage tissues. Chronic glucose overload not only increases the rate of oxidation and the generation of its harmful by-products but also the rate of glycation, a process whereby circulating glucose reacts and attaches to various proteins, such as hemoglobin and collagen. (Glycated hemoglobin is often used as a diagnostic tool and likely also underlies the association between metabolic dysregulation and anemia.) Circulating glucose can also react with metals, such as copper, becoming oxidized in the process. All of these processes lead to the formation of AGEs, which can inflict tissue damage by binding haphazardly to cell receptors, disrupting signaling cascades and inciting inflammatory responses.

The vascular endothelium seems especially vulnerable to the negative effects of excess glucose and AGEs in the bloodstream, showing signs of damage to the vascular walls well before any clinical markers of glucose dysregulation appear (Tooke and Hannemann 2000). Healthy endothelium has a smooth surface. Damage to the vascular walls sets the conditions for a potential range of cardiovascular-related disease. Microvascular pathologies, in which the small arteries are affected, appear tightly coupled with hyperglycemia; retinopathy (eye disease), neuropathy (nerve disease), and nephropathy (kidney disease) are all strongly associated with T2D (Nakagawa 2007). Notably, nephropathy is rarely seen in the absence of T2D. (Kidney dialysis is one of the most common and expensive treatments for T2D.) By the time symptoms appear, hyperglycemia has overwhelmed the kidneys and glucose has entered the urine. Glucose traveling through the renal tubules attracts water by osmosis, causing an increase in the frequency and volume of urination, often accompanied by thirst—the telltale signs of T2D.

Damage to the macrovascular system becomes evident in Syndrome X, characterized by hyperglycemia, hyperlipidemia, creeping hypertension, and the eventual appearance of atherosclerosis. Hypertension develops as damaged endothelial cells stimulate vascular remodeling, leading to a loss of vasodilation in the large arteries. Damage to the integrity of the endothelium also allows circulating lipids and LDLs to penetrate the arterial walls, where oxidation and inflammatory responses can ultimately lead to the formation of arterial plaque. Because the plaque core is thrombogenic (can rupture and trigger a blood clot), atherosclerosis increases the risk of heart attack and stroke. In addition, the steady accumulation of fat stores, which often accompanies metabolic dysfunction, tends to promote through the independent activity of adipokines a state of low-grade inflammation seen, in particular, in the arterial walls.

11.3.4.2   Brain Health

The systemic effects of metabolic dysfunction also appear to extend to the brain. T2D, for example, is a known risk factor for Alzheimer's Disease (AD), and AD patients have a higher risk of developing T2D (Correia et al. 2011). Although this association has been attributed to particular features of metabolic dysfunction, especially hyperglycemia, hypertension, and systemic inflammation (Cotman et al. 2007), two main differences between central and peripheral regulation raise the possibility that features common to T2D and AD may represent physiologic correlates, rather than direct causal pathways to cognitive impairment. First, the brain is highly selective about which substances can cross the blood–brain barrier, making the circulatory milieu in the brain distinct from the periphery. Second, the brain utilizes a unique set of complex pathways to deliver a steady supply of glucose to match demand. The hypothalamus, for example, has the capacity to "pull" glucose from the periphery, if necessary (such as during periods of psychological stress), by activating the sympathoadrenal system; catecholamines released from strategically placed nerve endings can interfere with both insulin production by beta cells and GLUT4 translocation in muscle and fat tissue (Peters 2011).

In my view, the linchpin linking Type-2 diabetes and Alzheimer's disease likely entails chronic physical inactivity. One of the essential mechanisms for maintaining brain function involves the interaction of three prominent growth factors: insulin-like growth factor-1 (IGF-1) produced by the liver, VEGF, and brain-derived neurotropic factor (BDNF). Levels of IGF-1 and VEGF increase in the periphery in response to physical activity and cross the blood–brain barrier to enter the brain. In the brain, the interactive effects of IGF-1 and VEGF induce the formation of new neurons and blood vessels, while the interactive effects of IGF-1 and BDNF serve to strengthen synapses (Cotman et al. 2007).

From an evolutionary perspective, it is not surprising that mounting evidence indicates that physical activity not only enhances learning and memory in all age groups, but also provides protective benefits against depression, Alzheimer's disease, and age-related cognitive decline (Ratey 2008; Brito 2009; Hwang and Kim 2015). Physical activity regulates the availability of key substrates necessary for brain function. Rather than emphasizing the clinical value of physical activity as an intervention strategy and remedy "uniquely positioned to improve brain health (Cotman et al. 2007, p. 469)," it may be worthwhile to see chronic inactivity as uniquely positioned to undermine brain health.

11.4 Conclusion

The field of evolutionary medicine posits that modern environments interact with ancestral physiology in ways that can promote disease. Here, I propose that a dysfunctional continuum emerges whenever a departure from ancestral conditions requires the human body (and psyche) to function in the equivalent of an extreme environment. Ascending to the highest point on Earth, the summit of Mount Everest (8848 m or 29,035 ft), becomes possible only because the human body utilizes mechanisms designed to respond to hypoxic conditions—honed over evolutionary time— encountered in the formative past at sea level, mainly in contexts involving physical exertion or injury and infection. The kidney, for example, shows a remarkable resilience to altitude hypoxia (maintains a high degree of functionality) because it was designed to withstand severe reductions in blood flow and, therefore, oxygen delivery during physical activity and exertion at sea level. At the same time, the brain is especially vulnerable to hypoxia because the brain's consumption of oxygen remains constant. (Balloonists found this out the hard way during flights in the early 1860s.)

Evolutionary biologists have pointed out for decades that many diseases of modernization, including metabolic disorders such as Type-2 diabetes, are largely preventable because they arise from contemporary lifestyle factors (Boyd Eaton et al. 1988; Fries et al. 1993; Boyd Eaton et al. 2002; Lindeberg 2012). Public health research continues to emphasize the role of prevention, showing evidence that investment in prevention is more cost effective, for example, than improved diabetes management, and the "considerable adverse impact of diabetes in terms of costs to society, health care systems, individuals and employers… increase[s] with diabetes duration as well as with the severity of the disease (Seuring et al. 2015, p. 825)." Medical prevention efforts often remain focused on better screening and diagnostic criteria, including for youth (Mancini 2009; Halpern et al. 2010; Canadian Task Force on Preventive Health Care 2012), and better managed-care models (Lee et al. 2011; Baptista et al. 2016). Improving diagnostic recommendations for metabolic disorders in the future based on the implications of the dysfunctional continuum framework presented here is a topic of ongoing research by the author.

Lieberman (2013, p. 351) asked in his concluding chapter, "How can an evolutionary perspective help chart a better future for the human body?" After identifying four possible approaches and eliminating three as essentially ineffective, Lieberman suggested that a government-regulated system of soft coercion or "nudges" could provide a promising path to prevention in alignment with evolutionary biology by changing the environment (approach 4), especially for children. Although relying on natural selection to sort the problem out (approach 1) or continuing to invest disproportionally in biomedical research and treatment (approach 2) hold little practical value for prevention, dismissing the efficacy of education and empowerment (approach 3) in favor of soft coercion may be premature—precisely because little is known about the efficacy of evolutionary biology knowledge itself as a prevention strategy. A small study I conducted with high school biology students in Cambridge, MA (Sherry, in prep) showed unequivocally that students who received instruction on diet and glucose regulation coupled with evolutionary biology shifted their perceptions about healthy food choices, compared to the control group of students who received instruction on diet and glucose regulation alone. Knowledge of evolutionary biology also led to precise dietary changes students intended to implement immediately, although follow-up studies would be needed to determine the extent of lasting behavioral change.

Given that health psychology and communication is a rapidly growing subfield in applied evolutionary medicine (Elton and O'Higgins 2008; Roberts 2011; Gibson and Lawson 2014), a conceptual framework based on evolutionary biology that identifies lifestyle behaviors as equivalent to entering an extreme environment conveys an immediate, intuitive sense of urgency and an implicit recognition that the human organism has strayed into a habitat for which it is ill equipped both physiologically and psychologically. Health risks associated with certain lifestyle factors, especially physical inactivity and dietary choices, may be more accurately perceived from this standpoint and effectively communicated as a public health message. From an evolutionary perspective, it makes sense why there is no inherent alarm system for major lifestyle departures from the formative ancestral past. At the same time, knowledge of evolutionary biology itself may provide an untapped yet promising tool for rendering the invisible extreme visible.

References

Baptista DR, Wiens A, Pontarolo R et al (2016) The chronic care model for type 2 diabetes: a systematic review. Diabetol Metab Syndr 8:7–13

Bateson G (1987) Steps to an ecology of mind: collected essays in anthropology, psychiatry, evolution and epistemology. Jason Aronson, Northvale

Beall CM, Laskowski D, Erzurum SC (2012) Nitric oxide in adaptation to altitude. Free Radic Biol Med 52:1123–1134

Boyd Eaton S, Shostak M, Konner M (1988) The Paleolithic prescription: a program of diet & exercise and a design for living. Harper & Row, New York

Boyd Eaton S, Strassman BI, Nesse RM et al (2002) Evolutionary health promotion. Prev Med 34:109–118

Bramble DM, Lieberman DE (2004) Endurance running and the evolution of *Homo*. Nature 432:345–352

Brito GNO (2009) Exercise and cognitive function: a hypothesis for the association of type II diabetes mellitus and Alzheimer's disease from an evolutionary perspective. Diabetol Metab Syndr 1:1–7

Canadian Task Force on Preventive Health Care (2012) Recommendations on screening for type 2 diabetes in adults. Can Med Assoc J 184(15):1687–1696

Correia SC, Santos RX, Perry G et al (2011) Insulin-resistant brain state: the culprit in sporadic Alzheimer's disease? Ageing Res Rev 10:264–273

Cotman CW, Berchtold NC, Christie L-A (2007) Exercise builds brain health: key roles of growth factor cascades and inflammation. Trends Neurosci 30:464–472

Dominiczak M (2007) Flesh and bones of metabolism. Elsevier, Philadelphia

Ellison PT (1990) Human ovarian function and reproductive ecology: new hypotheses. Am Anthropol 92:933–952

Ellison PT (2001) On fertile ground: a natural history of human reproduction. Harvard University Press, Cambridge

Elton S, O'Higgins P (eds) (2008) Medicine and evolution: current applications, future prospects. CRC Press, Boca Raton

Ewald PW (1994) Evolution and infectious disease. Oxford University Press, New York

Fries JF, Koop CE, Beadle CE et al (1993) Reducing health care costs by reducing the need and demand for medical services. N Engl J Med 329(5):321–325

Gibson MA, Lawson DW (eds) (2014) Applied evolutionary anthropology: Darwinian approaches to contemporary world issues. Springer, New York

Gluckman P, Hanson M (2006) Mismatch: the lifestyle diseases timebomb. Oxford University Press, Oxford

Greaves MF (2000) Cancer: the evolutionary legacy. Oxford University Press, New York

Halpern A, Mancini MC, Magalhães MEC et al (2010) Metabolic syndrome, dyslipidemia, hypertension and type 2 diabetes in youth: from diagnosis to treatment. Diabetol Metab Syndr 2:55–75

Hennige AM, Burks DJ, Ozcan U et al (2003) Upregulation of insulin receptor substrate-2 in pancreatic β cells prevents diabetes. J Clin Invest 112:1521–1532

Hwang HJ, Kim SH (2015) The association among three aspects of physical fitness and metabolic syndrome in a Korean elderly population. Diabetol Metab Syndr 7:112–117

Lee A, Siu CF, Leung KT et al (2011) General practice and social service partnership for better clinical outcomes, patient self efficacy and lifestyle behaviours of diabetic care: randomized control trial of a chronic care model. Postgrad Med J 87:688–693

Lieberman LS (2003) Dietary, evolutionary, and modernizing influences on the prevalence of type 2 diabetes. Annu Rev Nutr 23:345–377

Lieberman DE (2013) The story of the human body: evolution, health, and disease. Pantheon Books, New York

Lindeberg S (2012) Paleolithic diets as a model for prevention and treatment of western disease. Am J Hum Biol 24:110–115

Mancini MC (2009) Metabolic syndrome in children and adolescents—criteria for diagnosis. Diabetol Metab Syndr 1:20–23

Nakagawa T (2007) Uncoupling of the VEGF-endothelial nitric oxide axis in diabetic nephropathy: an explanation for the paradoxical effects of VEGF in renal disease. Am J Physiol Renal Physiol 292:F1665–F1672

Neal JM (2001) How the endocrine system works. Blackwell, Malden

Nesse RM, Williams GC (1994) Why we get sick: The new science of Darwinian medicine. Times Books, New York

Panter-Brick C, Worthman CM (eds) (1999) Hormones, health, and behavior: a socio-ecological and lifespan perspective. Cambridge University Press, Cambridge

Peters A (2011) The selfish brain: competition for energy resources. Am J Hum Biol 23:29–34

Petersen KF et al (2007) The role of skeletal muscle insulin resistance in the pathogenesis of the metabolic syndrome. Proc Natl Acad Sci U S A 104:12587–12594

Pollard TM (2008) Western diseases: an evolutionary perspective. Cambridge University Press, Cambridge

Pond CM (1998) The fats of life. Cambridge University Press, Cambridge

Rashidi A, Kirkwood TBL, Shanley DP (2009) Metabolic evolution suggests an explanation for the weakness of antioxidant defenses in beta-cells. Mech Ageing Dev 130:216–221

Ratey JJ (2008) Spark: the revolutionary new science of exercise and the brain. Little, Brown and Company, New York

Roberts C (ed) (2011) Applied evolutionary psychology. Oxford University Press, Oxford

Seuring T, Archangelidi O, Suhrcke M (2015) The economic costs of type 2 diabetes: a global systematic review. Pharmacoeconomics 33:811–831

Sherry DS, Marlowe FW (2007) Anthropometric data indicate nutritional homogeneity in Hadza foragers. Am J Hum Biol 19:107–116

Stearns SC, Koella JC (eds) (2008) Evolution in health and disease. Oxford University Press, New York

Stearns SC, Nesse RM, Govindaraju DR, Ellison PT (2010) Evolution in health and medicine Sackler colloquium: evolutionary perspectives on health and medicine. Proc Natl Acad Sci 107(l):1691–1695

Tooke JE, Hannemann MM (2000) Adverse endothelial function and the insulin resistance syndrome. J Intern Med 247:425–431

Trevathan WR, Smith EO, McKenna JJ (eds) (2008) Evolutionary medicine and health: new perspectives. Oxford University Press, New York

Wells JCK (2010) The evolutionary biology of human body fatness: thrift and control. Cambridge University Press, Cambridge

West JB, Schoene RB, Luks AM, Milledge JS (eds) (2013) High altitude medicine and physiology, 5th edn. CRC, Boca Raton

Willingham DT (2009) Why don't students like school? A cognitive scientist answers questions about how the mind works and what it means for your classroom. Jossey-Bass, San Francisco

Index

© Springer Science+Business Media New York 2017
G. Jasienska et al. (eds.), *The Arc of Life*, DOI 10.1007/978-1-4939-4038-7

Printed in the United States
By Bookmasters